The Language
of **Technical**
Communication

Ray Gallon

The Content
Wrangler
Content Strategy Series

The Language of Technical Communication

Copyright © 2016 The Content Wrangler, Inc.

Credits

Series Producer:	Scott Abel
Indexer:	Cheryl Landes
Series Cover Designer:	Marc Posch
Publishing Advisor:	Don Day
Publisher:	Richard Hamilton

Disclaimer

The information in this book is provided on an "as is" basis, without warranty. While every effort has been taken by the authors and XML Press in the preparation of this book, the authors and XML Press shall have neither liability nor responsibility to any person or entity with respect to any loss or damages arising from the information contained herein.

This book contains links to third-party websites that are not under the control of the authors or XML Press. The authors and XML Press are not responsible for the content of any linked site. Inclusion of a link in this book does not imply that the authors or XML Press endorse or accept any responsibility for the content of that third-party site.

Trademarks

XML Press and the XML Press logo are trademarks of XML Press.

All terms mentioned in this book that are known to be trademarks or service marks have been capitalized as appropriate. Use of a term in this book should not be regarded as affecting the validity of any trademark or service mark.

XML Press
Laguna Hills, California
http://xmlpress.net

First Edition
ISBN: 978-1-937434-48-9 (print)
ISBN: 978-1-937434-49-6 (ebook)

This book is dedicated to the memory of Ellen Stewart,
whose prophecy took thirty years to come true.
If she were here, she'd be rocking back and forth
in rhythm to her broom, sweeping the floor of her theatre,
smiling an eldritch smile, and saying, "I told you so."

Table of Contents

3

Foreword

The field of technical communication has been around as long as people have used processes, procedures, products, and services. But recently, the field has been evolving more and more quickly, as new technologies and new ways to reach our audiences have become available.

I recently said to my wife, "I linked our Nest with Alexa." I didn't need to explain that I had linked our Google Nest thermostat with our Amazon Echo voice-recognition device, allowing us to control our home temperature with voice commands. This sentence, which would have been complete nonsense five years ago, is perfectly meaningful now, because the *Internet of Things* has become ubiquitous in our daily lives.

We are continually reminded of the power of devices that sense and interpret context. My phone knows where I am. It knows when I'm leaving work to drive home and reports my expected commute time. It knows what sports teams I follow and what companies I like, and it knows when I'm traveling.

These technologies grant amazing new capabilities for understanding and reaching our audiences. The best customer experiences occur when documentation anticipates needs, context, and goals, and these technologies, many of which are discussed in this book, make it easier for us to create those kinds of customer experiences.

The terms defined in this book range from basic terminology that most of us know to the latest concepts pushing the boundaries of the discipline. Do you need a refresher in basic and invariant technical communication concepts? Consider the Core Concepts section. Want to dive deeper? The Technical Concepts section will ground you in approaches for creating, managing, and delivering your content using proven processes.

The Standards and Conventions section will bring you up-to-date on the latest standards-based approaches, which will help you manage your content more effectively, especially in the face of changing tools and technologies.

Would *controlled language* and *structured authoring* help to improve the consistency of your content and decrease translation costs? Would *topic-based authoring* help to improve the readability of your content? Are you collecting and making appropriate use of *metadata*? What is *content*? Why is *localization* not enough? Why should technical communicators also have *business analysis* skills?

How are you and your organization responding to changes in the outside world? Take a look at the Future Directions section, where you can read about topics such as *wearables, augmented reality,* and *artificial intelligence.*

How do you deliver content? The Deliverable Presentations section may provide new ideas for serving an increasingly mobile audience that expects information on demand, in the appropriate form, targeted for their situation, and available on any device. This audience also expects information to be more interactive, to actively respond to context, and, as always, to address their needs in the most concise, direct manner possible.

As technical communicators, we seek a user-centered approach that provides the right information, in the right way, at the right time to make someone's life easier and more productive. Part of our tool kit for implementing a user-centered approach is the vocabulary we use with our colleagues.

If we have a common understanding of the language of technical communication, we can be more effective and efficient in our interactions with our colleagues. Whether you are a newly hired trainee or a seasoned professional, understanding the body of terms in this book will make it easier to work together.

While our methods and techniques may be changing, the mission of technical communication remains unchanged: explaining processes, procedures, products, and services. And helping make people successful. This book can help you be more effective in carrying out that mission.

Alan Houser
Group Wellesley, Inc.
Consultant and Trainer, Technical Publishing
http://groupwellesley.com

Preface

As I sat down to write this preface, I had an epiphany: here we have a book of technical communication terminology, and yet the term *technical communication* is nowhere defined in it! Then I had a second thought: isn't that so like us? I've been in this profession for over 20 years, and I still have family who ask, "So, I've never been able to figure out exactly what you do, can you explain it?"

During my tenure on the board of directors of the Society for Technical Communication (STC) we made a decision to stop trying to define the *profession* and refer to technical communication as a *discipline*. I think this is the right course. So herewith, I present to you, dear readers, a selected list of terms pertinent to the discipline that we all practice, but that I am not going to define – I'll leave it to you to answer your own families' curiosity in your own way.

This book is an example of the joy and the terror of being a technical communicator. We are a varied, multi-talented lot, and we do a huge variety of things, many of which are not in our job descriptions. This means that any list of terms will necessarily contain some exotic notions, and leave out some terms that might be considered essential. Our choices were dictated by two overarching admonitions: that the list not be limited to technical *writing*, but include all aspects of technical communication; and that the list continue to be valid and applicable into the foreseeable future. Following these prescriptions was challenging and produced at least one editorial crisis, just when we thought we had it all figured out. I believe this book is the better for it.

We are an international team. To start with, Richard and I successfully collaborated, using the nine-hour time difference between California and Europe to keep some tasks going 'round the clock. The contributors also come from both sides of the Atlantic, and represent practitioners in a variety of roles and industries, from different linguistic and cultural backgrounds, an illustration of just how global our discipline is.

This book is also an example of how we practice what we preach. We wrote the terms in a wiki built on a specialized DITA schema, and we practice reuse. Those of you who have read *The Language of Content Strategy*, another book in this series, will recognize certain terms: content, accessibility, findability, metadata, single sourcing, structured content, content reuse, transclusion, intelligent content, augmented reality, governance model, localization, globalization, and terminology management were all found there. In some cases, they are reused as is, and in others, they have been modified to be more directly applicable to technical communication. In one case, the technology is changing so fast that we wrote a whole new definition. There is so much overlap between content strategy and technical communication that we have also provided an appendix in the form of a glossary, which contains more content strategy terms that we think are essential for a technical communicator.

The job of editor of a collective work is made vastly more enjoyable and – dare I say it – easier when there are first-class contributors. And here, I have to say that we have been more than fortunate. Most of the credit for any success this book may have goes to the 52 experts who authored terms, with sincere gratitude for their skill, their willingness to rewrite and be edited, and the quality of their work.

Having Richard Hamilton as a sidekick has also made this book a joyful process, in both his roles as fellow editor and as publisher. Don Day worked tirelessly, sometimes day and night, to get our wiki up and running, and his openness to our suggestions and his willingness to roll with the punches has also facilitated the cause enormously. And special thanks to Scott Abel, The Content Wrangler, for the opportunity to create this book and for precipitating the editorial crisis that has made it so much better.

All of us offer this book to you in the hope that it will serve you long and well and help to enlarge your horizons.

Ray Gallon
The Transformation Society
http://transformationsociety.net

Core Concepts

The Core Concepts section covers the foundational terms that underpin the discipline of technical communication. Building a common understanding of these terms and their importance enhances our ability to communicate effectively with other practitioners.

- Content
- Document
- Intelligent Content
- User Assistance
- User Experience
- Project Management
- Business Analysis
- Governance Model
- Metadata
- Transclusion
- Responsive Design
- Instructional Design
- Interaction
- Usability
- Accessibility
- Findability
- Globalization
- Localization
- Indexing

Scott Abel
Content

What is it?
Any text, image, video, decoration, or user-consumable elements that contribute to comprehension.

Why is it important?
Content is the single most-used way of understanding an organization's products or services, stories, and brand.

About Scott Abel
Known affectionately as "The Content Wrangler," Scott Abel is an internationally recognized global content strategist and intelligent content evangelist who specializes in helping organizations deliver the right content to the right audience, anywhere, anytime, on any device.

Scott is the founder, CEO, and chief strategist at The Content Wrangler, Inc. He's also a highly sought after keynote presenter, moderator, and a frequent contributor to content industry publications. Scott's alter ego, The Audio Wrangler, is a popular DJ and dance music mashup artist.

Email	scottabel@mac.com
Website	thecontentwrangler.com/
Twitter	@scottabel
LinkedIn	linkedin.com/in/scottabel
Facebook	facebook.com/scottpatrickabel

Why does a technical communicator need to know this?

Content can be described in several ways, some technical, others conceptual:

- **Contextualized data:** Data is a context-free value; content has enough context to aid with consumer comprehension. For example, the number "12" is merely data. Adding context to the data, such as 12th month or 12 years old, imbues the data with meaning and creates content.
- **The stuff inside a container**: In a world where content is virtually always touched by technology, this means content is between a set of standardized markup tags, allowing technology to automate the processing of content.
- **An extension of the user experience:** Content is the treasure at the end of the treasure hunt. Without good content, the best user experience falls flat.

Most importantly, *content is a business asset* that we use to communicate with our customers, prospects, and investors. Content is how we communicate our brand, how we acquire and retain customers, how we drive our reputations, and how we build a social enterprise. It is the lifeblood of any organization.

To develop effective content strategies, it is important to understand both the editorial and technical sides of content.

- Editorially, content should be relevant, accurate, informative, timely, and engaging and should conform to editorial standards.
- Technically, content should be standards-based, use well-formed schemas, be semantically rich for filtering and findability, and be structured to support automated delivery.

This allows content to be converged, integrated, and syndicated – all important aspects of leveraging content to its fullest potential.

Robert J. Glushko
Document

What is it?

A self-contained unit of information, most often with text-intensive content in printed or digital formats. In its broadest sense, a document can also include audio, video, or interactive content as well as text. We can even stretch this definition to fit documents that are enhanced with sensing, computational, and communications capabilities, such as augmented reality documents that overlay context-sensitive information on their base content.

Why is it important?

Documents organize the interactions between people, between people and enterprises and institutions, and between the machines and computers that are ubiquitous and essential in today's information-intensive world.

About Robert J. Glushko

Robert J. Glushko is an Adjunct Full Professor at the University of California, Berkeley's School of Information, where he has been since 2002. He has over thirty years of R&D, consulting, and entrepreneurial experience in information systems and service design, information architecture, content management, electronic publishing, Internet commerce, and human factors in computing systems.

He founded or co-founded four companies, including Veo Systems in 1997, which pioneered the use of XML for electronic business before its 1999 acquisition by Commerce One.

Email	glushko@berkeley.edu
Website	people.ischool.berkeley.edu/~glushko/
Twitter	@rjglushko

Why does a technical communicator need to know this?

Every important advance in business, manufacturing, communications, finance, government, and other spheres of human activity have required new kinds of information exchanges with specialized types of documents. The idea of documents as self-contained units of information that satisfy the requirements for information exchange is a technology-independent abstraction perfectly suited for a world bursting with an exponentially growing number of computers, communication devices, sensor-enabled smart things, and other producers and consumers of documents.

Technical communicators can easily understand the logical separation of document content, format, and appearance that is possible with digital documents, but the historical function of documents as evidence reinforces the idea that a document's format or technology implementation is inherently part of it. It is not always easy to maintain this separation in practice. This might seem like a philosophical debate, but it can have significant business, legal, and technological implications.

Technical communicators are often the people best suited for designing documents because they focus on the people who use them and the purposes for which they are used. Technical communicators often assist document engineers in analyzing the information requirements for the documents that define the inputs and outputs, or the requests and responses, that make up the logical model for a document type. Technical communicators typically design the presentations or renditions of these document models that enable their effective creation, use and comprehension by people.

Ann Rockley
Intelligent Content

What is it?

Content that is modular, structured, reusable, format-free, and semantically rich and, as a consequence, discoverable, reconfigurable, and adaptable.

Why is it important?

Intelligent content enables organizations to rapidly adapt their content to the changing needs of their customers and the devices they use.

About Ann Rockley

Ann Rockley is CEO of The Rockley Group. She is a pioneer in content reuse, intelligent content strategies for multi-channel delivery, and content management best practices. Known as the "mother" of content strategy, she introduced the concept with her best-selling book, *Managing Enterprise Content: A Unified Content Strategy, 2nd edition* (Peachpit, 2012).

Email	rockley@rockley.com
Website	rockley.com
Twitter	@arockley
LinkedIn	linkedin.com/in/annrockleytrg

Why does a technical communicator need to know this?

In today's multichannel environment, we have to move away from the artisanal creation of content to a manufacturing model in which consistently structured, reusable components can be assembled into a variety of deliverables. Getting the correct content to the right customer in the right context and on the device of their choosing requires intelligent content.

An intelligent content strategy establishes a coherent plan under which content can be designed, developed, and delivered to achieve maximum benefit to the customer and organization while minimizing costs. It provides measurable ROI through better use of finite resources, reduced costs at every phase of content development, increased quality, and unlimited device delivery.

Intelligent content is supported by:

- **Content models:** define the structure of content components and assemblies, formalized in templates, forms, and markup languages like XML. This format-free, style-less content can be automatically displayed, filtered, or layered to optimize display on a target device with little or no human intervention. Writing guidelines help authors write consistently structured content.
- **A reuse strategy:** identifies what types of content will be reused, the level of granularity, how content will be reused, and how to automatically assemble reusable content. With modular, reusable content, you can change the order of components, include or exclude components, and reuse components to build entirely new types of content to meet new needs.
- **A taxonomy strategy:** defines how to store and retrieve your content based on a common vocabulary (*metadata*). Using metadata, you can retrieve the pieces of content you need to automatically build customized information sets.

Ellis Pratt
User Assistance

What is it?

A broad term for describing the different forms of content that can help people use a software application or other technological product correctly.

Why is it important?

User assistance places the work that technical communicators do within the framework of user-centered design, of improving the experience of a user, rather than seeing that work as simply creating manuals.

About Ellis Pratt

Ellis Pratt is a Director at Cherryleaf, a technical writing services and training company based near London, in the UK. He has over 20 years' experience working in the field of documentation. Ellis is also on the management council for the Institute of Scientific and Technical Communicators (ISTC).

Email	ellis@cherryleaf.com
Website	cherryleaf.com
Twitter	@ellispratt
LinkedIn	uk.linkedin.com/in/ellispratt

Why does a technical communicator need to know this?

It's very easy to think about what technical communicators do by looking at what they produce. 50 years ago, you'd have heard "they're the people who write manuals," and more recently, it might have been "they're the people who write Windows Help." If technical communication is seen as only creating specific items, there's a danger that technical communicators will miss out on opportunities to do things differently.

Indeed, new technologies have led technical communicators to look for an appropriate term for describing all the different deliverables that can be provided today. User assistance fills that role.

This enables technical communicators to talk about the process as well as the deliverables. Instead of saying, "let's write a Help file," we ask "what is the best possible way to help the user for this situation?" User assistance can include: text in the user interface, online help, *wizards*, "What's This?" help (text that appears when you move your mouse over a word), on-boarding screens, videos, software tutorials, quick reference cards, or PDF/printed manuals. It also includes information embedded in the application itself – preempting problems users might experience by making information more accessible and more relevant.

Delivering good user assistance still requires the ability to write clearly and succinctly. However, it also requires an understanding of the technologies that can be used to deliver content and an understanding of your users and their needs.

Noz Urbina
User Experience

What is it?

The overall outcome of a user's interaction with a product, service, or organization across all touch-points, experienced through the lens of the user's personal background and preferences.

Why is it important?

User experience is the comprehensive, holistic measurement of organizational performance and understanding of the market, because every user is unique and any link in the chain can harm their experience.

About Noz Urbina

Noz Urbina is a globally recognized content strategist and modeler. He's well known as a pioneer in adaptive content modeling to support rich, personalized, and contextually relevant content across the user journey. He is also co-author of the book *Content Strategy: Connecting the dots between business, brand, and benefits* (XML Press, 2012).

Email	noz.urbina@urbinaconsulting.com
Website	urbinaconsulting.com
Twitter	@nozurbina
LinkedIn	es.linkedin.com/in/bnozurbina
Facebook	facebook.com/noz.urbina.3

Why does a technical communicator need to know this?

If you want to create a great experience for your users when they come to your content, you have to understand their particular situation. A technical communication team must understand its users, and their relationship with products and services, across the entire lifecycle of that relationship.

Technical communicators need to be empowered to develop a clear understanding of user personas and meaningful metrics for tracking user experience (UX), both of which are necessary to evaluate the performance of content in a user-centric way. In other words, if you're not measuring user experience, you're only measuring for yourself.

All communications and customer-facing staff need to participate in delivering experiences. From websites to packaging to event collateral, from mailers to presentations to user interface text, from manuals to answers read over the phone or delivered by instant message: all communications are threads linking the user and the organization. These threads intertwine to sew the tapestry of user experience.

Focusing on UX will help you uncover new ways to deliver your content, for example, embedding it in the user interface, delivering interactive portals, or even creating wearable tech or augmented reality. Being user-centric helps you get out of your own mindset and limitations and rethink content from the outside in.

Delivering value-added content as a contribution to UX generates positive *content karma*: the more value you deliver to your users, the more value will come back your way. When each touch-point is optimized to work in concert, users will be more positive and engaged.

Erin Vang
Project Management

What is it?

Everything needed to complete (or end) a temporary effort, preferably with schedule, budget, scope, quality, and risk constraints in balance.

Why is it important?

Writing well is just the start. You also have to organize and deliver your work on time and on budget, in a way that leaves your collaborators willing to work with you again.

About Erin Vang

Erin Vang, PMP, has several decades of experience in commercial software documentation, quality assurance, project and program management, localization, content strategy, and people management, currently as a technical communication manager at Dolby Laboratories.

Previously she worked in R&D for the JMP division at SAS, at Abacus Concepts, and SYSTAT, and in 2008 she formed the consultancy Global Pragmatica LLC®, offering services in facilitative leadership, localization, and custom statistical tool development.

Email	erin.vang@globalpragmatica.com
Website	globalpragmatica.com
Twitter	@erinvang
LinkedIn	linkedin.com/in/erinvang
Facebook	facebook.com/erin.vang

Why does a technical communicator need to know this?

Understanding something and explaining it is the core responsibility of a technical communicator. However, getting to do that job more than once requires taking more responsibility than that. Sustainable success requires planning your work effectively and communicating realistically with all your stakeholders about what is important to each of them.

Writing projects – like all projects – rarely go exactly as planned. Scope, schedule, and budget tend to change once projects are underway, and managing your stakeholders' expectations about that is part of the job. You must be able to recognize when the plan needs to change and enroll stakeholders in change management, for example, by calculating and securing their agreement to an adjusted schedule and budget. You must be able to anticipate risks and enroll stakeholders in appropriate risk management. Finally – getting back to the core responsibility – you must actually deliver the expected results on time and within budget, as defined by the agreed-upon adjusted plan.

Often, writers serve not just as resources on project teams but also as leaders within those teams. The lead or sole writer on a project is often responsible for planning and managing relevant tasks within that project. Approaches vary, but writers are generally expected to plan and monitor their tasks and dependencies. Whether projects use waterfall-style work breakdown schedules or newer methodologies such as agile development, where scrum team members work in sprints to complete stories on a project backlog, writers have both resource and leadership roles to play.

Good writers employ their verbal and illustrative communication skills, their bigger-picture views of user experience design, and their customer-advocacy attitudes to increase team velocity both directly and indirectly, and, as a result, they often find themselves recruited into project and people management roles.

A good writer doesn't just write well but also organizes the project of writing something well.

Christopher Ward
Business Analysis

What is it?

A research discipline that provides strategic solutions to business needs by analyzing changing markets and industry trends.

Why is it important?

Business analysis helps technical communicators better understand the audience they are writing for and helps them align content goals with strategic business goals, thus increasing product value and consumer loyalty.

About Christopher Ward

Christopher Ward, Director of Sales at WebWorks, specializes in helping teams accomplish big things by better aligning departmental processes with overall company strategies. Christopher's experience in strategy development began as an Analyst for U.S. Army Intelligence. Later, he moved to the corporate world with Dell computers. His diverse experiences allow Christopher to recognize untapped potential in a company's overall business strategy and help them achieve that potential. He has traveled all over the world delivering the message that "Documentation is not a business expense, it is a revenue generating tool."

Email	christopher@webworks.com
Website	webworks.com
Twitter	@WebWorksChris
LinkedIn	linkedin.com/in/webworkschris
Facebook	facebook.com/WebWorksChris

Why does a technical communicator need to know this?

Companies spend large amounts of money on business analysis. Yet too many technical communicators are unaware of this valuable information. The audience for technical content is the same as the market your company serves. Understanding business analysis can help technical communicators create informative and engaging content that aligns with business goals. These elements must be accounted for to create successful content, making technical communication some of the most difficult content to author.

Business analysis provides a real-time analysis of the changing market (consumer base/audience) and can give technical communicators much-needed information to help them meet the requirements of business content creation.

Solutions are becoming more complex in today's business environment, and they have to be easier for a consumer to use. This leads businesses to either put huge demands on their technical communicators or abandon them all together. Business analysis can help technical communicators understand not only the solution a company delivers, but also what specific features about that solution are most important to particular markets. Technical communicators can also use business analysis metrics to determine market reaction to their content. This supports process improvements and increases the value of the product or service and the business.

Finally, business analysis can help us recognize trends in a changing market. Technical communicators can use this information to predict the future needs of their market and write to them. This marriage of content creation and business analysis will become the most pivotal element of a business' continued success.

Val Swisher
Governance Model

What is it?
Guidelines that determine who has ownership and responsibility for various aspects of an organization.

Why is it important?
A governance model makes it clear who has authority to make which decisions about content and contributes to smooth operational decisions and processes.

About Val Swisher
Val Swisher, CEO of Content Rules, is an expert in global content strategy, content development, and terminology management. Val helps companies solve complex content problems by analyzing their content and how it is created. Her customers include Google, Facebook, GoPro, Rockwell Automation, and others.

Val is the author of *Global Content Strategy: A Primer* (XML Press, 2014).

Email	vals@contentrules.com
Website	contentrules.com
Twitter	@valswisher
LinkedIn	linkedin.com/in/valswisher

Why does a technical communicator need to know this?

Having a governance model is a critical aspect of implementing and maintaining a successful technical documentation department. It is important to have a clear chain of command and a clear understanding about who is responsible for which decisions.

Often, technical documentation organizations do not have a governance model. People in the organization assume that everyone knows who the decision makers are along the content lifecycle. Unfortunately, this is usually not the case.

When the governance model is not clearly defined, the technical documentation workflow can become stuck and unable to move forward. This is because the decision maker(s) are either unaware of the decisions that need to be made, think that someone else is responsible for making the decisions, or do not understand the urgency of their role.

By creating a governance model, organizations can ensure that workflows and decision trees are clearly delineated, agreed to by all parties, and strictly adhered to.

Laura Creekmore
Metadata

What is it?

Attributes of content you can use to structure, semantically define, and target content.

Why is it important?

Metadata extends the capabilities of content, making it more powerful and effecting efficient operation in a data-driven world.

About Laura Creekmore

Laura Creekmore and her company, Creek Content, develop content strategy and information architecture for organizations with complex communication needs. She teaches content strategy as an adjunct faculty member at Kent State University and serves as the associate editor for IA for the ASIS&T Bulletin.

Email	laura@creekcontent.com
Website	creekcontent.com
Twitter	@lauracreekmore
LinkedIn	linkedin.com/in/lauracreekmore

Why does a technical communicator need to know this?

Metadata can take many forms. In its simplest form, metadata describes attributes or constraints of a content field, providing additional information that tells software how to handle content. You can use metadata in conjunction with business rules to design and deploy your content more effectively.

For example, you can use metadata to represent how content relates to one or more subjects, and then use those relationships on a website to create links between topics and help content. The metadata helps you create links on the fly and in context, and it helps ensure that the links remain up to date when content changes.

Metadata describes content, but not how that content should be displayed. Imagine the confusion if your metadata included the option to tag a video as "featured." What happens when you tag more than one video in that way? What happens if you forget which one you tagged before? It's much better for your metadata to include dates, times, subjects, content types, and constraints, then you can build display and presentation rules that *use* that metadata to determine how, when, and under what conditions that content should be displayed.

A metadata standard like *Dublin Core* may simplify your work and make your content more extensible. However, many people create their own metadata, customized for internal use.

Metadata is sometimes revealed to users (in *faceted search*, for instance), but most often, it's the behind-the-scenes workhorse that makes your life easier and gives systems context about your content to make it more powerful.

Eliot Kimber
Transclusion

What is it?

The inclusion of content from one source into another source by hyper-link reference. The presented result appears as though the included content had occurred at the point of reference.

Why is it important?

First formalized as the idea of link-based use-by-reference, transclusion is a fundamental feature for any content representation system, such as DITA, that enables true reuse.

About Eliot Kimber

Eliot has been working with structured markup in one form or another for over 30 years. Eliot's career focus has been on the authoring, management, and production of large-scale hyperdocuments, ranging from the Networking Systems library at IBM in the '90s to complex DITA-based publications today.

Eliot is the author of *DITA for Practitioners, Volume 1: Architecture and Technology* (XML Press, 2012).

Email	ekimber@contrext.com
Website	contrext.com
Twitter	@drmacro
LinkedIn	linkedin.com/in/eliotkimber

Why does a technical communicator need to know this?

The term *transclusion* is important to technical communicators because, as a single term, it embodies the challenging concept of *use-by-reference* as implemented (or not really implemented) by many content authoring systems. Technical communicators may see the term in technical discussions around markup language design, system design, and system implementation. The term is also used by *information architects* and document system engineers.

Technical communicators should understand the history of the term. Transclusion is an essential part of most hypertext systems. The term was coined by hypertext pioneer Ted Nelson in his seminal work *Literary Machines* (Mindful Press, 1994), a book that influenced all hypertext systems and approaches that followed it.

Some XML-based applications support built-in transclusion features; examples include DITA's content reference (conref) facility and the W3C XInclude facility used by DocBook. It is the rare technical communicator who does not use some form of transclusion.

Systems that support transclusion also bring unavoidable complexity in terms of authoring effort, hyperlink management, configuration management, quality assurance of reused content, and so on. Technical communicators need to understand the cost that comes with the power of transclusion so they can choose when to accept those costs. It is easy to transclude yourself into serious trouble. However, used correctly, transclusion enables more efficient authoring and management of information.

Charles Cooper
Responsive Design

What is it?
A design technique that allows content on a web page to automatically reflow, resize, reformat, and reposition itself so it can be displayed to its best advantage on a variety of device sizes and orientations.

Why is it important?
Our content is consumed on mobile devices of all shapes and sizes. Responsive design can help us display our content well without needing to reformat it for various devices and screen sizes.

About Charles Cooper
Charles Cooper is VP of The Rockley Group. He's been involved in creating and testing digital content for more than 20 years. He's passionate about user experience, taxonomy, workflow design, composition, digital publishing and mobile delivery. Charles teaches, facilitates modeling sessions, and develops taxonomy and workflow strategies for clients around the world.

Charles is co-author of the best-selling book, *Managing Enterprise Content: A Unified Content Strategy, 2ⁿᵈ edition* (Peachpit, 2012), and *Intelligent Content: A Primer* (XML Press, 2015).

Email	cooper@rockley.com
Website	rockley.com
Twitter	@Cooper42
LinkedIn	linkedin.com/in/charlescoopertrg

Why does a technical communicator need to know this?

We once created technical content for specific page sizes. When we started writing for the web, we continued using that technique, creating fixed-size designs that worked well on traditional desktop screens. That no longer works. People experience our content on a wide variety of devices with different screen sizes, aspect ratios, and orientations.

We need to think outside of the box, whether that box is 640×480, 1920×1080, or 144×168. With today's devices, that box might be round, as on a smart watch. It might even be totally out of the box, floating in air using augmented reality techniques.

As a technical communicator, you probably won't be writing code to implement responsive design, but you do need to know how it affects you. You must think differently about the content, the needs of your users and how you create that content.

Some textual content may not be visible or usable on some devices. Images will resize, appear in different locations, or in some cases not appear at all. Many of the techniques we use without thinking (such as referring to a figure by saying, "see the image below") just aren't applicable or can't be used reliably.

Rather than writing for a large screen or page and then extracting information to be used on a small device, it's often better to write for a small device and use *progressive disclosure* techniques to expand content as required.

Responsive design is an exciting way of thinking; we need to recognize the challenges and embrace the opportunities.

Phylise Banner
Instructional Design

What is it?
The process of analyzing learner needs in alignment with desired learning outcomes, followed by the development of learning environments through the management of content, interaction, and assessment in support of those learning outcomes.

Why is it important?
Instructional design drives the development of quality learning experiences, addressing how information is imparted to the learner, how the learner interacts with that content, and to what extent desired goals and outcomes have been met.

About Phylise Banner
Phylise Banner is the Director of Online Teaching and Learning at Union Graduate College, where she is working to expand online offerings through a mindful institution-wide approach to program, course, and faculty development in alignment with teaching and learning effective practices and the Community of Inquiry framework.

An Adobe Education Leader and STC Fellow, Phylise has been working in the field of online teaching and learning since 1997, planning, designing, developing, and delivering online courses, programs, and faculty development initiatives.

Email   pbanner@gmail.com
Website   phylisebanner.com
Twitter   @phylisebanner
LinkedIn   linkedin.com/in/phylisebanner

Why does a technical communicator need to know this?

Instructional design addresses how material is taught. Focusing on effective, efficient, and appealing content, the instructional design process provides a means to engage and immerse the learner in the learning experience.

Technical communicators are regularly tasked with creating learning materials, including content, interaction, and assessment components. Instructional design enables technical communicators to understand how learners will approach their materials and to create robust learning environments that support established goals and desired outcomes.

Instructional design promotes effective planning in the alignment of content, interaction, and assessment with standards and objectives. Technical communicators employ these instructional design principles and practices to facilitate knowledge acquisition and to assess understanding and mastery.

User assistance designers and developers can benefit from incorporating instructional design strategies into their workflow by implementing the learning-needs-focused practices modeled throughout the instructional design process.

Key to effective technical communication is the consistent review of content and the user experience. The instructional design process can be used to model this reflective process by guiding the regular evaluation of the effectiveness of instructional materials and the practices through which learning is taking place.

Gerry McGovern
Interaction

What is it?

The process by which a person uses a technology to communicate or perform tasks together with a computer, mobile device, or other technological product.

Why is it important?

More and more, people are interacting with technologies in similar ways as with other humans. This interaction with technology has become part of our daily, routine conduct as we live our lives and get things done.

About Gerry McGovern

Gerry McGovern helps large organizations become more customer centric on the web. His commercial clients include Microsoft, Cisco, NetApp, VMware, and IBM. He has also consulted with the US, UK, Dutch, Canadian, Norwegian and Irish governments. He is the founder and CEO of Customer Carewords, a company that has developed a set of tools and methods to help large organizations identify and optimize their customers' top online tasks. He has written five books on how the web has facilitated the rise of customer power. The Irish Times described Gerry as one of five visionaries who have had a major impact on the development of the web. In 2015, he was shortlisted for a Webby for his writings.

Email	gerry@customercarewords.com
Website	customercarewords.com
Twitter	@gerrymcgovern

Why does a technical communicator need to know this?

Interaction is about use, and customers judge technology based on how easy it is to use. We need to focus not only on what we write or create, but especially on how what we have created is used. How easy is it for customers to interact with what we have created? Does it help them solve their problems? Do they get the right answer?

The new interactive model forces technical communicators to think about the customer experience and the outcome – the way our words are used. The customer is active when addressing technical information. That information might be embedded in a software interface or in a website; it can appear as an interactive eBook or overlaid in an augmented reality application. These interactive systems are a far cry from static textual manuals. They communicate with the user following the same rules and principles as human communication.

This means that a technical communicator needs to take into account not only content, voice, and tone, but also the relationship of customers to the product – what are they trying to do and how will they know if they're successful?

To measure interaction, we must measure use. We must come up with a series of customer tasks and then get real customers to use the technology to try and complete these tasks. We measure two key things: success rate and time on task. Our job is to support the interaction and help customers complete their tasks as quickly and easily as possible.

Eric Reiss
Usability

What is it?

The degree to which an individual can accomplish specific tasks and achieve broader goals while using a particular tool or service.

Why is it important?

Despite the constant overuse of the term and misuse of the research, industry professionals have long known that good usability often holds the key to business success both on- and off-line.

About Eric Reiss

Eric Reiss has held a wide range of jobs including: piano player, senior copywriter, player-piano repairman, jukebox restorer, pool hustler, school-bus driver, cartoonist, magician, adventure-game creator, starving student, showboat actor, and stage director. The breadth and depth of his experience have served him admirably as a usability professional.

Today, Eric is CEO of the FatDUX Group in Copenhagen, Denmark, an international user-experience design company with offices and associates in over a dozen cities worldwide. He also has several books to his credit, including the best-selling *Usable Usability* (John Wiley, 2012).

Email	er@fatdux.com
Website	fatdux.com
Twitter	@elreiss
LinkedIn	linkedin.com/in/ericreiss
Facebook	facebook.com/eric.l.reiss

Why does a technical communicator need to know this?

Usability is traditionally associated with the technical or functional aspects of an interactive product – ease of use. For example, do the buttons work? Are they conveniently placed? Is the server response time fast enough, etc. But technical communicators need to focus on the psychological side of the usability coin: elegance and clarity. We need to communicate in a clear, concise, and understandable manner that creates a shared frame of reference with the reader.

This is the list I give my "Writing for the Web" students:

- Don't take anything for granted
- Anticipate the questions people might have
- Answer the questions they didn't think to ask
- Examine your content (words, images, sounds) in the context of your reader's situation

As always, the communication environment – the time and place surrounding an experience – will affect the nature of the information needed (or provided) at any given time. For example, reading a printed manual when setting up a new smart TV will be very a different experience from that of an engineer looking for troubleshooting tips on a smart phone somewhere out in the field.

And finally (and perhaps most important of all), remember:

- Whatever you say, say it clearly, without resorting to insider terms
- Don't assume everyone reads as carefully as you write

Char James-Tanny
Accessibility

What is it?

The extent to which content is available, understandable, and usable by all, regardless of disabilities or impairments such as sensory, physical, cognitive, intellectual, or situational.

Why is it important?

Accessibility equals usability for (almost) everyone. Many people think, "Oh, for the blind," when accessibility is mentioned, but it encompasses much more than that.

About Char James-Tanny

Currently working in Technical Publications at Schneider Electric, Char has over 35 years of experience as a technical communicator. She has spoken around the world about accessibility, social media, web standards, collaboration, and technology. She belongs to the Boston Accessibility Group and is the Primary Coordinator of the Boston Accessibility Event.

Email	char.james-tanny@schneider-electric.com
Twitter	@CharJTF
LinkedIn	linkedin.com/in/charjtf
Facebook	facebook.com/CharJTF

Why does a technical communicator need to know this?

One of the main tenets of technical communication is to *know your audience*. Unless you create content that is available, understandable, and usable, the chances are good that you're ignoring as much as 20% of your audience.

How can you make content more available to people with disabilities? Accessibility happens during design, development, and delivery. Many content strategy best practices already address accessibility:

- Use descriptive headings (formatted with tags or styles).
- Use short sentences (fewer than 25 words) and short paragraphs (no more than three sentences).
- Write in second person, active voice, and present tense.
- Use the best word, not the longest.
- Use descriptive hyperlink text. For example, use "Learn how to create a title" as the link.
- Avoid location words, such as *earlier*, *later*, or *below*.
- Include keystrokes when possible. For example: **Select File → New or press ALT+F, N.**

Take these additional steps to create accessible formatting and markup:

- Use styles. Never manually format text so that it looks the way you want.
- Left-justify text for left-to-right languages and right-justify for right-to-left languages.
- Use the correct color contrast (3:1 for large text and 4.5:1 for other text and images).
- Use a font size comfortable for almost everyone (at least 12 points).
- Restrict the number of font families to three.
- Add the alt attribute to images (unless they're only decorative).
- Format lists as lists. (Don't use a symbol to indicate a bullet or type the number.)

Making your content more accessible helps everyone, not just people with disabilities.

Cheryl Landes
Findability

What is it?

The quality of being able to discover and retrieve technical information through searching or browsing.

Why is it important?

Technical information must be found to be consumed; otherwise, it is useless.

About Cheryl Landes

Cheryl Landes founded Tabby Cat Communications in Seattle in 1995. She has more than 25 years of experience as a technical communicator in computer software, HVAC/energy savings, marine transportation, retail, and manufacturing. She specializes as a findability strategist, helping businesses organize content to flow logically and make content easier to retrieve.

Email	clandes407@aol.com
Website	tabbycatco.com
Twitter	@landesc
LinkedIn	linkedin.com/in/clandes

Why does a technical communicator need to know this?

Finding information is the biggest business challenge. According to recent surveys, employees spend 30% of their workdays looking for information. They often make eight searches before they find the correct information (Cottrill Research, *Various Survey Statistics: Workers Spend Too Much Time Searching for Information*).

Your organization can have the best technical documentation: simple and well-crafted. But if search engines cannot understand your content or users can't find it, then no one will ever get to read that information. If users can't find your content, they'll look elsewhere.

Users find information with a combination of two intentional methods: navigation and search:

- **Navigation** means using the available options and contextual clues to locate technical content. Buttons, tabs, tables of contents, menus, links, and indexes are common navigation options online. In print, readers navigate with the help of tables of content, indices, and page numbers.
- **Search** is the act of looking for specific content by entering a query in a search engine or application. Users form queries with keywords or search terms, and the search engine displays the query results based on its index and understanding of the request.

Increasing the findability of technical information means ensuring that content has metadata and structure that enables search engines and consumers to locate and retrieve relevant content, as well as making it more likely that users will encounter the information they need. Well-designed navigation and content designed for search help serve up the information users want, when they want it. Headings and subheadings that clearly describe what the technical content is about also help users scan content easily and locate the details they need quickly.

Bill Swallow
Globalization

What is it?
The analysis of, and planning for, the development, delivery, and consumption of global content; in essence, globalization is the analysis that forms a global content strategy.

Why is it important?
Globalization reveals the benefits, risks, needs, and demands of content among all target consumers and influences better decision-making for global information exchange.

About Bill Swallow
Bill specializes in content strategy with an emphasis on handling challenging localization and terminology scenarios. His experience on both the client and vendor side of localization has made him (rather painfully) aware of all of the potential pitfalls in content localization.

Email	bswallow@scriptorium.com
Website	scriptorium.com
Twitter	@billswallow
LinkedIn	linkedin.com/in/billswallow
Facebook	facebook.com/techcommdood

Why does a technical communicator need to know this?

Globalization, sometimes abbreviated as g11n, is a term traditionally grounded in economics. As we continue to reach across geographic and cultural boundaries, we need to be increasingly mindful of how other cultures affect, and are affected by, our efforts. Intended or not, cross-cultural exchange is a mutual integration into a much larger system of both mutual and conflicting ideas, beliefs, expectations, and demands. In the case of content, globalization is the study of these factors for the betterment of information exchange.

Content globalization looks at the entire lifecycle of content, from inception to destruction, with heavy consideration for multicultural needs and demands. When performing this analysis, it is best to work backwards from the consumption of content, considering the audience and its needs.

From there, similarities and differences can be drawn based on appropriateness (both cultural and legal), necessity, form, and timeliness. These form the necessary requirements that inform the strategic engineering of the entire content development process, from style to technology to methodology to even who the writers are.

While globalization is traditionally regarded as an effect to be measured and studied, the proliferation of buzzwords in the global content arena (translation, localization, and internationalization, for example) indirectly suggests that globalization is an act or process. We can certainly approach it as such, so long as we are using the analysis of the effect (whether actual within existing contexts or theoretical when considering new ones) to influence positive changes in, and outcomes from, our global content strategy.

James V. Romano
Localization

What is it?
Adaptation of content to make it more meaningful, appropriate, and effective for a particular culture, locale, or market.

Why is it important?
Localization increases the relevance of content for a particular target audience.

About James V. Romano
Ever since he was a small child trying to understand his Italian grand-father at the dinner table, James Romano has been trying to unravel the mysteries of languages and cultures. For thirty years now, his company, Prisma International, has been helping clients communicate with their global customers, audiences, and users.

Email   jromano@prisma.com
Website   prisma.com

Why does a technical communicator need to know this?

Localization, sometimes abbreviated as l10n, is an essential process for technical communicators. More than just an add-on at the end of the content development cycle, localization requires careful planning and strategy right from the start.

To be effective, content must be relevant and meaningful to the target audience. Localization is the process by which content is made more appropriate and more meaningful for a particular culture. Without localization, technical communicators would be spinning generic pablum in the hope that users (near or far) will recognize a scintilla of meaning, latch on, and perhaps buy the product, heed the warning, or swallow the pitch.

Localization is strategic: it requires a comprehensive, planned approach in which all parts of the content system – the messaging, technologies, audience – come together in a dynamic, creative process, producing what can best be described as an "aha moment." Localized content taps into the power of local culture and uses it to project and amplify its message to create a deeper, more resonant message.

Localization is about producing an *aha* in any language, culture, or medium. Content that doesn't lead to an *aha* falls flat. On the other hand, a localization-driven content strategy is capable of producing meaningful content experiences for its audience(s), creating *aha*s in Anchorage, Andorra, and Anhui.

In sum, technical communicators need to know how to leverage the power of culture to create successful content experiences, and localization is the means by which they can harness that power.

Jan Wright
Indexing

What is it?

A set of organized, easily-navigated, and concise terms and phrases linked to locations in content, giving users fast access.

Why is it important?

Indexes link concepts and coordinates with metadata, providing users with a bottom-up tool for navigating content, crossing author-created boundaries such as chapters or topics, and democratizing all concepts for easy retrieval.

About Jan Wright

Jan Wright specializes in embedded indexing, single-sourcing, and technical indexing. Her background includes print production, InDesign, and interactive and ebook indexes. She co-chaired the American Society for Indexing's Digital Trends Task Force, focusing on linked eBook indexes. Honors include the ASI/H.W. Wilson Award for Excellence in Indexing and ASI's Hines Award for her service to the Society.

Email	jancw@wrightinformation.com
Website	wrightinformation.com
Twitter	@windexing
LinkedIn	linkedin.com/in/JanCWright
Facebook	facebook.com/Wright-Information-Indexing-Services-235865426486044/

Why does a technical communicator need to know this?

Without access, content is unusable. Accessible content provides a variety of navigation tools, reflecting the many ways users search for information. Search and tables of contents are the most widely used navigational tools, but indexes support capabilities that make them equally important. These capabilities include visibility, scan-ability, support for alternative terms, and cross references to related terms. Indexes allow users to browse through subtopics, alternate phrasing hints, and content structure. They support a wide variety of readers, including new readers, readers who are unsure whether their topic is covered, readers who are returning for known topics, and readers who are unsure what terms to use.

You can think of indexing as a double-headed fork with tines on both ends. The left side of the fork represents alternate terms (tines), which lead to the concept's best terminology. The fork's handle links that term to the right-hand tines, which represent result locations in the content. Left tines become the terms in your index. They are often organized from A–Z, but they can be sorted chronologically, geographically, symbolically, or with imagery. Right tines (locators) become the coordinates that point to results, which may reside in print, digital formats, audio, video, or 3D space.

To create user-friendly structures, indexers gather audience information and survey the content. They predict most-likely searches and write headings, pulling important concepts into the alphabetized stream. Names, synonyms, and labels are disambiguated. Expert or beginner terminology can be emphasized. Competitor terms, slang, acronyms, and topic breakdowns provide browseable clues. Indexers then create index metadata, which can be embedded as codes in content or maintained in separate, standalone files with location links.

Index data is used to display merged indexes, ranges, or micro-indexes – also known as metadata subsets. Index data can also be used to boost search results.

Indexing provides users with a browseable method for researching questions, links content in an easy-to-understand navigation scheme, and tags content at a micro-level with metadata that can be coordinated with tables of contents, menus, taxonomies, folksonomies, or search engines.

Technical Concepts

This section covers terminology related to the technology that supports technical communication. This ranges from methodologies, such as topic-based authoring, to systems, such as learning management systems, to enabling technologies, such as conditional content. While technical communicators may not need to know how to implement these technical concepts, they should understand how to use them in their work.

- Structured Content
- Topic-Based Authoring
- Content Reuse
- Single Sourcing
- Content Architecture
- Component Content Management System
- Learning Management System (LMS)
- Conditional Content
- Content Variables
- Dynamic Delivery

Don Day
Structured Content

What is it?
Content, whether in a textual, visual, or playable format, that conforms to structural and semantic rules that allow machine processing to meet specific business requirements.

Why is it important?
Structure in a document involves identifying the scope and relationship of meaningful parts. Named structures enable both logical processing and independent styling of what readers see.

About Don Day

Don Day is a content engineer with deep experience with innovative authoring solutions and information architectures for structured, semantic content for the web and across the enterprise.

He provides consulting on strategy, technology, and best practices for optimizing the value and usefulness of unstructured data. Don created the expeDITA platform, which was used to develop the content for this book.

Email	donday@learningbywrote.com
Website	learningbywrote.com
Twitter	@donrday
LinkedIn	linkedin.com/in/donrday/
Facebook	facebook.com/donrday

Why does a technical communicator need to know this?

Humans are much better than computers when it comes to understanding the nuances of content. Readers generally understand the visual intent of style in what they read in a browser or in print. But if a computer needs to know that intent as well (as in, "search for all warnings in the install guide"), it helps to somehow indicate the scope and meaning in a computer-readable manner.

By indicating the order and intent of the parts of a document, writers ensure that publishing tools well into the future can usefully process that content, even if reading technologies change.

Adding structure to content adds both present and future value, turning content from a single-use commodity into a long-term asset. Content can be structured in a number of ways. The most common ways are to apply semantic markup, such as *XML (Extensible Markup Language)*, or to store content in named fields in a database.

Structured content clearly indicates not only the parts of the discourse (the titles, sections, lists, tables, and phrases that represent organization) but also the semantic intent of those containers. For example, paragraphs identified more specifically as quotations can be not only rendered differently for readers, but also made more easily discovered in searches for quotations or citations.

By structuring content appropriately, you can more easily turn information into knowledge, instructions into automation, concepts into lesson units, and more, thereby increasing its value to the business.

Mark Baker
Topic-Based Authoring

What is it?

Authoring an information set as a collection of discrete units called *topics*, rather than as a whole book or help system.

Why is it important?

Readers are increasingly information-snacking on small pieces of content which they find by searching, and small discrete units of information can be produced and managed more efficiently.

About Mark Baker

Mark Baker is the author of *Every Page is Page One: Topic-based Writing for Technical Communication and the Web* and the forthcoming *Structured Writing: Principles and Practices*, both from XML Press. He offers training and consulting on topic-based authoring through his company, Analecta Communications Inc.

Email	mbaker@analecta.com
Website	analecta.com
Twitter	@mbakeranalecta
LinkedIn	ca.linkedin.com/in/mbakeranalecta

Why does a technical communicator need to know this?

Readers today typically search large information sets or the Internet when they have a technical problem, looking for the right piece of information. How easy it is to Google a product is now a common selection criteria for technology. A traditional manual is both too small to search and too big to read. Therefore, technical writers increasingly write topics rather than books.

Modern readers also expect information to be kept perpetually current. Topic-based information sets support rapid updates as topics can be verified and published individually.

A topic is a small, self-sufficient unit of information. Unlike an article, a topic does not stand entirely alone, but is connected to a larger information set. Some topic-based information sets are hypertexts, similar to Wikipedia. Some are organized hierarchically, others like a database. Some are static and some are built dynamically. Some are still organized as traditional manuals.

The word *topic* covers both the unit that is written and the unit that is read. In some cases, however, the authored topic may be smaller than the topic that is presented to the reader. The reader's topic is then composed of multiple authored topics. This typically occurs where content reuse is a major business concern.

Topics frequently conform to a specific pattern or type, which helps readers recognize particular topics as the type of information they are looking for and navigate more easily. Conforming to types also helps writers be more complete and consistent and may make content easier to reuse.

Kristen James Eberlein
Content Reuse

What is it?
The practice of using content components in multiple information products.

Why is it important?
Developing reusable content that can be used in multiple places and output formats saves valuable resources, enforces consistency, and improves content quality and effectiveness.

About Kristen James Eberlein
Kristen James Eberlein is an information architect who works with clients that use the Darwin Information Typing Architecture (DITA). She chairs the OASIS Technical Committee that develops the DITA standard.

Email   kris@eberleinconsulting.com
Website   eberleinconsulting.com
Twitter   @kriseberlein

Why does a technical communicator need to know this?

Content reuse is a key tactical component of a content strategy. Efficient content reuse enables single sourcing and multi-channel publishing; enforces editorial consistency; conserves time and fiscal resources; and can help ensure accurate, compliant (and thus effective) content.

Efficient content reuse does not involve copy-and-pasting; instead it uses *transclusion*, whereby content is authored in one location and used by reference in other locations. Many Extensible Markup Language (XML) architectures implement transclusion; perhaps the most well-known is the Darwin Information Typing Architecture (DITA). Many authoring systems and content management systems also include proprietary mechanisms for transclusion.

Companies can maximize content reuse by developing structured content that is standards-based and semantically rich. Content can be reused at different levels of granularity:

- An entire information product
- An entire topic or collections of topics
- Elements of a topic

In addition, content can be designed so that conditional processing (filtering) can generate different variants of information products. A content analysis can determine the appropriate level of granularity. A reuse strategy should define the method of content reuse, what content should be reused, the granularity of reuse, how reused content is controlled, and who owns reused content.

You can effectively manage reused content by employing a content management system (CMS) to control access, determine where the controlled content is used, and identify potentially reusable content. When content is structured well, content managers can employ automation to power content reuse, for example by pre-populating information products with reused content or using tools such as to help prevent content authors from accidentally deleting the controlled content.

Leigh W. White
Single Sourcing

What is it?

Creating content once, planning for its reuse in multiple places, contexts, and output channels.

Why is it important?

Single sourcing enables authors to leverage content to its fullest potential, with benefits such as increased consistency and accuracy and reduced development time.

About Leigh W. White

Leigh White is a DITA author and information architect. As a DITA Specialist at IXIASOFT, she helps documentation groups transition into the IXIASOFT DITA CMS.

Leigh is the author of *DITA For Print: A DITA Open Toolkit Workbook* (XML Press, 2013) and a contributor to *The Language of Content Strategy* (XML Press, 2014).

Email	leigh.white@ixiasoft.com
Website	ixiasoft.com
Twitter	@leighww
LinkedIn	linkedin.com/in/leighwwhite

Why does a technical communicator need to know this?

Single sourcing is an approach to developing content that can be used to produce multiple outputs in different formats for different platforms. With this approach, content creators only need to maintain one set of source content, greatly reducing authoring, editing, and translation time, as well as reducing the risk of introducing inconsistencies between multiple, often redundant, content sets.

One key to single sourcing is separating content from formatting. Rather than formatting the content while authoring, the content is formatted as part of the publishing process. This frees authors to concentrate solely on the quality of the content and allows designers to format content appropriately for each channel. Single-sourced content is usually in an open format, such as XML, which describes the content semantically so it can be processed intelligently based on the nature of the information and its intended use.

Successful single sourcing requires a solid plan for content creation and content reuse. The two go hand-in-hand. When creating content, authors must be mindful of all the ways in which it might be used. It's up to content-strategy-minded technical communicators to develop the plan for intelligent reuse.

Content creators must follow architectural guidelines when writing content to ensure its maximum reusability in multiple contexts. The content must be sufficiently granular, and it must share a common voice and vocabulary. For single sourcing to succeed, content creators must collaborate with information architects and end users to regularly re-evaluate content in its various uses.

Marie Girard
Content Architecture

What is it?
The definition and organization of pieces of information (content) so that their use is consistent, logical, and efficient.

Why is it important?
Content architecture enhances the business value of content by designing a scalable structure that supports content strategy, improves user experience, and facilitates the work of technical communicators.

About Marie Girard
Marie Girard manages content strategy and architecture for IBM products. She leads unified content strategy efforts through collaboration across silos, content audits, and metrics. When Marie is not busy taming content chaos at IBM, she teaches technical communication at Paris Diderot University and keeps investigating how everything interrelates through the practice of yoga.

Email	magichopi@openmailbox.org
Website	about.me/magichopi
Twitter	@MarieGirardChop
LinkedIn	fr.linkedin.com/in/girardmarie

Why does a technical communicator need to know this?

When a content project involves multiple authors, stakeholders, and users, content architecture becomes necessary. While information architecture is a discipline of user experience, content architecture encompasses the wider challenges of content modeling, author experience, workflows, and technological constraints. Content architecture brings consistency, logic, and efficiency to complex content systems.

- **Consistency:** Through the design of models, templates, and guidelines, content architecture supports coherent organization and style in a content system. It helps implement structured information models based on XML, and it may complement them by defining controlled vocabularies, labeling guidelines, and stylistic and visual recommendations for overall consistency.
- **Logic:** Content architecture ensures that the various pieces of content fit together to create a larger picture, spotting redundancies and contradictions. A general workflow shapes how users access and navigate content and how technical authors create and retrieve it. This workflow integrates text as well as more interactive types of content to provide the right type of media at the right time.
- **Efficiency:** Content architecture identifies common patterns and structures to help find content and reuse it. Because a good author experience supports a good user experience, content architecture includes the design of sustainable content storage structures that work with content management systems and help manage reuse. These structures also take globalization into account, improving translation management.

Content architecture is a holistic approach to organizing content systems, which considers not only content, but also people, processes, and technology.

Nolwenn Kerzreho
Component Content Management System

What is it?

A centralized system that helps organizations capture, manage, store, preserve, and deliver topic-based content (components), whether the content is proprietary or follows a standard architecture, like DITA.

Why is it important?

A Component Content Management System (CCMS) provides a working environment for content engineers and technical communicators to plan, track, reuse, publish, translate, and control topic-based content assets.

About Nolwenn Kerzreho

As the Technical Account Manager for IXIASOFT, Nolwenn Kerzreho helps customers realize the benefits of structured content, cross-silo collaboration, and DITA content management. An associate teacher at Rennes University, Nolwenn has a keen eye on the evolution of skills requirements for the next generation of technical writers.

Email	nolwenn.kerzreho@ixiasoft.com
Website	ixiasoft.com
Twitter	@nolwennIXIASOFT
LinkedIn	fr.linkedin.com/in/nolwennkerzreho

Why does a technical communicator need to know this?

A CCMS is a content management system designed to handle the challenges that come with granular content. It provides adapted ways to find smaller pieces of content using full-text and metadata searches, variables, branches, and so on. It also tracks dependencies, reuse (transclusion) between objects, references to resource files, and content applicability.

A CCMS enables single-sourcing for high-quality, fully-automated, and multi-channel publishing and can manage all content assets included in the final information products, including XML fragments, artwork, lists, videos, and 3D models.

A CCMS provides:

- **Centralized Control:** concentrates knowledge and enables sharing, swapping, reusing, and reviewing content. It can answer content audit questions such as, who did what, when and why did they do it, on what content did they work, and where was that content delivered.
- **Content and Media Enrichment:** helps create metadata that facilitate discovery by both CCMS users and information product users.
- **Referential Integrity:** manages and maintains the links between objects – particularly important when the content is reused in multiple documents, versions, and formats.
- **Cross-functional Collaboration:** integrates and adapts to the organization's roles and access needs, facilitating experts' collaboration in adding and reviewing content.
- **Process Management:** adapts and enforces the organization's business rules and processes, including planning, translation monitoring, final processing, and output delivery.
- **Process Integration:** interfaces with delivery platforms, such as dynamic delivery systems, learning management systems, and other storage sources.

As more organizations shift towards using topic-based content strategies, technical communicators will find themselves working within CCMS environments and appreciating the support provided by such systems.

Kevin Siegel
Learning Management System (LMS)

What is it?

A complex website that allows you to track learner access to your eLearning content, set time limits, run quiz pass/fail reports, and automatically award completion certificates.

Why is it important?

While eLearning content can be created easily with off-the-shelf-tools, you will need a Learning Management System (LMS) to deliver your content and to help you gauge its effectiveness.

About Kevin Siegel

Kevin Siegel is the founder and president of IconLogic, Inc. He has written hundreds of books on applications ranging from Adobe Captivate to Articulate Storyline. Kevin spent five years in the U.S. Coast Guard as an award-winning photojournalist and has three decades experience as a print publisher, technical communicator, instructional designer, script writer, and eLearning developer. He is a certified technical trainer, a veteran classroom instructor, and a frequent keynote speaker and facilitator at conferences all over the world.

Email	ksiegel@iconlogic.com
Website	iconlogic.com
Twitter	@Kevin_Siegel
LinkedIn	linkedin.com/in/kevinsiegel
Facebook	facebook.com/IconLogicInc

Why does a technical communicator need to know this?

Technical communicators often develop eLearning content. This includes content written specifically for training and content re-purposed for training. In either case, you may need to interact with Learning Management Systems (LMS).

A Learning Management System manages the entire workflow for delivering training. It controls access, delivers your content to learners, controls the order in which material is presented, manages testing, tracks progress, and prepares status reports. In short, an LMS is to eLearning content what a Component Content Management System is to technical content, plus it acts as a publishing engine and wiki, too.

There are many choices of LMSs that will handle delivery and management of eLearning content. Solutions range in price from free (Moodle is arguably the most well-known open-source LMS) to millions of dollars. Most of today's LMSs, from the least to the most expensive, support eLearning content created by modern eLearning tools.

Julio Vazquez
Conditional Content

What is it?
Content that has sufficient metadata to allow a processor to filter or flag that content in any output format, using a profile to determine the exact output for a given context or format.

Why is it important?
Conditional content facilitates the reuse of content components in multiple contexts or formats. The metadata specifies the contexts to which a specific component applies.

About Julio Vazquez
Julio Vazquez is a Sr. Content Analyst with Vasont Systems. He has been involved in technical communication for over 30 years and was part of the IBM team that developed the Darwin Information Typing Architecture (DITA). He is the author of *Practical DITA* and has written articles about DITA and information architecture.

Email	jvazquez@vasont.com
Website	vasont.com
Twitter	@juliov27612
LinkedIn	linkedin.com/in/juliovazquez

Why does a technical communicator need to know this?

More than ever, technical communicators are required to do more with less. Meeting requirements for ever-shortening development cycles with fewer resources requires detailed analysis of all information being published. That analysis should identify opportunities to use similar content in different contexts.

Conditional content includes all the variations of content within a single topic, so the writer only has one topic to manage for all contexts. A processor uses a profile to determine which portions of the topic become part of the output. Using conditional content can minimize the number of topics the writing team has to manage.

Consider a topic that lists the features of a family of cell phones. The basic features of all the cell phones are the base topic. Additional or different features are added to the base and given attribute values the processor uses to include or exclude content based on the product manual required for a particular phone.

Another example uses conditional content to support a device that does not render tables optimally. The topic would contain a presentation specific to that device, then when preparing content for that device, the processor removes the table and includes the alternate presentation instead.

Resist the temptation to create alternative texts within the structure of a sentence or for a topic that contains little base content and becomes just a series of filtered paragraphs or sentences. These structures become difficult to maintain and can cause problems for localization.

Nancy Harrison
Content Variables

What is it?

Variables that contain phrase-level content that needs to be in a topic no matter what document the topic is part of, but that changes depending on context, for example, a product name or a company name.

Why is it important?

Content variables have been a key factor in allowing reuse of content across products and platforms. By isolating terms or phrases that are likely both to appear in multiple places and to change depending on factors external to the content itself, those terms or phrases can be modified for publication without modifying the actual topic content.

About Nancy Harrison

Nancy Harrison has many years of experience as a technical writer, documentation manager, and information architect. She is an active contributor to the XML, DITA (currently DITA TC secretary), and DocBook standards. Her company, Infobridge Solutions, provides consulting in the fields of information architecture, DITA strategy, and stylesheet design.

Email	nharrison@infobridge-solutions.com
Website	infobridge-solutions.com
Twitter	@nancylph
LinkedIn	linkedin.com/in/nancyharrison1
Facebook	facebook.com/nancy.harrison.1650

Why does a technical communicator need to know this?

In technical writing, certain pieces of information (for example, product or company names) appear repeatedly and may be changed arbitrarily by people or circumstances beyond your control. If you enter this information as text, you need to perform a time-consuming search-and-replace operation any time it gets changed, and if any instance has a typo, you won't find it, and it won't get changed.

If you identify such information and use content variables to encapsulate it everywhere it's used, then you only need to change it in one place when it is changed, and your publishing environment will produce the right output. In addition, if you use the same content for multiple products, distributed simultaneously, feature names or product names may need to change depending on which product the content is being used with. In this case, content variables and a dynamic publishing environment can enable you to simultaneously publish documentation for multiple products.

To take advantage of existing variables and avoid duplication, you need to be able to identify chunks of content that should be placed in content variables and discover what content variables are already being used within your organization. If your organization is not currently using content variables, you might want to investigate how you can incorporate them into your writing process to improve efficiency and consistency.

Chris Despopoulos
Dynamic Delivery

What is it?

The assembly of content *after* receiving a request, so the system can filter or merge different sources, process the results, and return content that is relevant to you at the moment you make the request.

Why is it important?

Given cloud and virtual technology, software systems are increasingly dynamic. The reader is also increasingly dynamic, whether using different devices or filling different roles. Static delivery simply can't keep pace. Dynamic delivery captures the current states of system and reader and returns content that is specific and meaningful.

About Chris Despopoulos

Chris Despopoulos has been documenting software systems and implementing publishing tools for the last 25 years. His coding experience informs his writing, and his writing experience informs the designs of his publishing tools.

Most recently he's been perfecting 4D Pubs; Distributed Dynamic Document Display that implements content assembly in the client. 4D Pubs is the help delivery system used by VMTurbo, the demand-driven control platform for virtual IT.

Email   despopoulos_chriss@yahoo.com
LinkedIn   linkedin.com/in/chris-despopoulos-3544457

Why does a technical communicator need to know this?

Technical writing as a profession is constantly evolving. Each new wave of technology brings new requirements and possibilities into the mainstream. Recent developments in virtual, cloud, and container technologies make the network more fluid than ever. Microservice architecture changes the concept of application, giving us constellations of services that can be different for different people.

Consider the *Internet of Things (IoT)*. No two people will have the same set of things, and a system comprising a unique set of things will exhibit unique behavior. Dynamic delivery can assemble content that is relevant to the current system, in the system's current state, according to the profile of the person or system making the request.

Static content addresses system variations by hard-coding reuse and filtering criteria. These techniques can support only so much complexity – trying to hard code every possible combination doesn't scale. Further, if you can't predict the full set of components at any moment, then you can't hard code your criteria.

As systems become more complex, we will have no choice but to let machines calculate the different combinations. Technical writers will need to understand how this works, the new capabilities it brings, and how to write topics that play well in a dynamic environment.

Standards and Conventions

The discipline of technical communication is built on a foundation of standards, including standard concepts, such as *minimalist information design* and *controlled language*, and internationally-recognized standards, such as the family of XML standards. Even previously closed formats are becoming open standards supported by organizations such as OASIS, the World Wide Web Consortium (W3C), and the International Organization for Standardization (ISO).

Industry standards offer technical communicators easier content interchange, better assurance that content can be migrated to new software without losing information or becoming tied to an obsolete platform, and a broader range of tools to manage and deliver content.

- Minimalist Information Design
- Omnichannel
- XML Document Editing Standards
- Media Standards for XML
- XML Processors
- eLearning and mLearning Standards
- Terminology Management
- Controlled Language
- European Machinery Directive

John M. Carroll
Minimalist Information Design

What is it?

Designing information to evoke, guide, sustain, and leverage human action.

Why is it important?

Minimalist information design is important because, in contexts of engaged activity, people neither want, nor can effectively use, comprehensive information.

About John M. Carroll

John M. Carroll is Distinguished Professor of Information Sciences and Technology at the Pennsylvania State University. He received the Rigo Award and the CHI Lifetime Achievement Award from ACM, and the Goldsmith Award from IEEE.

He is a fellow of AAAS, ACM, IEEE, and the Association for Psychological Science, and honorary fellow of the Human Factors and Ergonomics Society and the Society for Technical Communication. In 2012, he received an honorary doctorate in engineering from Universidad Carlos III de Madrid.

Email   jcarroll@ist.psu.edu
Website   jcarroll.ist.psu.edu/

Why does a technical communicator need to know this?

Technical communicators often design information resources to support engaged activity contexts, in which people are attempting to carry out tasks, to improvise based on their prior understanding, and to learn by doing.

Examples include human-computer interaction and use of mobile/personal devices. In such activity contexts, the tolerance of people for information that is not immediately relevant and actionable is quite low.

Minimalist information design addresses this challenge by focusing on the specific goals people will want to pursue or may need to pursue (error recovery). Minimalist information resources invite people to act and help them identify appropriate goals and concrete actions they can take to progress toward their goals. Minimalist information helps and encourages people to make sense of and reflect upon their own interactions in the course of planning and carrying out those interactions.

Rahel Anne Bailie
Omnichannel

What is it?

A marketing approach which endeavors to provide customers with a seamless user experience, no matter through which channel or device, or during which stage of the content, product, or customer lifecycle.

Why is it important?

Successful omnichannel marketing depends on delivering the content appropriate to their stage in the lifecycle through the most appropriate channels.

About Rahel Anne Bailie

Rahel Anne Bailie is Chief Knowledge Officer at Scroll, an STC Fellow, an industry author, and a results-driven content strategist. She has a strong track record of developing successful digital content projects, tackling the complexities of managing content for clients globally. Her strength is diagnostics: calculating how to use content to deliver compelling experiences.

Rahel is co-author of *Content Strategy: Connecting the dots between business, brand, and benefits* (XML Press, 2013) and co-editor of *The Language of Content Strategy* (XML Press, 2014).

Email	rahel.bailie@gmail.com
Website	scroll.co.uk
Twitter	@rahelab
LinkedIn	linkedin.com/in/rahelannebailie
Facebook	facebook.com/public/Rahel-Anne-Bailie

Why does a technical communicator need to know this?

Omnichannel, in the context of technical communication, is the recognition that customers consume content in various ways, in various forms and times during the customer lifecycle. Creating unique content for every medium (print and online), channel (marketing brochures, data sheets, specification sheets), knowledge base (information hubs, print catalogues, online catalogues), and so on would be prohibitively expensive and a maintenance nightmare.

The way for technical communicators to work in an omnichannel environment is through multichannel publishing, where a superset of content is stored with the multiple variations tagged for use toward omnichannel efforts, often driven by marketing but critical to producing technical communication.

The content can be segmented by audience, market, language, product, product version, information type, output medium, or any number of variations. Content can be delivered to downstream systems, ready to be used in the appropriate way. This not only ensures that the important details are consistent across channels, it also means that maintenance is a more efficient and reliable process.

A simplified version of how a superset of product content feeds an omnichannel initiative might look like this:

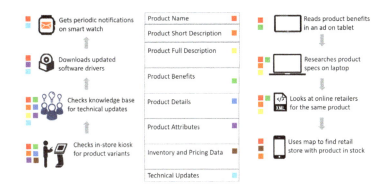

Norman Walsh
XML Document Editing Standards

What is it?

Vocabularies and processes used to create specialized, editable structures for documents that conform to the XML standard.

Why is it important?

Successful interchange of structured information depends on a common understanding of the vocabularies involved and how they are processed. The standards for providing this take the form of an XML schema and provide common ways of structuring different types of content to facilitate reuse, catalog maintenance, version control, and other aspects of technical documents.

About Norman Walsh

Norman Walsh is a Lead Engineer at MarkLogic Corporation where he helps to develop APIs and tools for the world's leading enterprise NoSQL database. He is also an active participant in a number of standards efforts worldwide: he is chair of the XML Processing Model Working Group at the W3C where he is also co-chair of the XML Core Working Group. At OASIS, he is chair of the DocBook Technical Committee.

With two decades of industry experience, Norm is well known for his work on DocBook and a wide range of open source projects. He is the author of *DocBook 5: The Definitive Guide* (O'Reilly Media, 2010).

Email	ndw@nwalsh.com
Website	nwalsh.com/
Twitter	@ndw
LinkedIn	linkedin.com/in/normanwalsh

Why does a technical communicator need to know this?

There are many ways to store and process structured information. Within the realm of prose communications, as distinct from data-interchange APIs, XML is one of the most common and successful ways to do so. Most publishing workflows rely on XML at some point by, for example, reading the underlying XML structures of Microsoft Word .docx files as part of a conversion process, generating XML markup for an automated publishing system, or publishing eBooks.

XML document editing standards identify common vocabularies and best practices for reducing costs and increasing reliability. Technical communicators are most likely to encounter one of these XML editing standards:

- **DITA (Darwin Information Typing Architecture):** originally developed by IBM, DITA is now maintained as an open standard by OASIS. DITA is based on a topic model, but it is useful for most types of technical documentation. DITA is notable for specialization, a capability that allows developers to extend the vocabulary while remaining compliant with the DITA standard.
- **DocBook:** the most mature standard on this list, DocBook was originally designed and implemented by HaL Computer Systems and O'Reilly & Associates around 1991. Today, it is maintained by OASIS. Although originally designed around a book model, it has evolved to be useful for help systems, ebooks, and other forms of technical documentation. DocBook is widely used in the open source world.
- **oManual:** started when O'Reilly Media and the iFixit online repair website wanted a data format for exchanging their procedural manuals. It is a standard for semantic, multimedia-rich procedural manuals and also an API designed to provide quick access from shared locations (e.g. cloud) on mobile and other devices. It is an open standard, maintained by oManual.org, and has been approved as an IEEE standard.
- **S1000D:** originally developed by the AeroSpace and Defense Industries Association of Europe (ASD) for use with military aircraft. It is now also used in many other parts-intensive contexts. It is maintained by the S1000D Steering Committee, which is made up of aerospace and defense representatives from around the world.

Understanding and applying XML document editing standards reduces the risk of misunderstanding, leverages existing tools and technologies, reduces costs, and increases productivity.

Scott Hudson
Media Standards for XML

What is it?

Industry-defined file formats for images, video, audio, and other multi-sensory content that can be included as a resource in a structured XML document.

Why is it important?

Media standards provide a consistent technical method for delivering these experiences to a variety of output formats from single-source content (XML). The prominent documentation standards have dedicated elements for referencing media: DocBook has `<inlinemediaobject>` and `<mediaobject>` elements; DITA has `<image>` and `<object>`; TEI has the `<media>` element.

About Scott Hudson

Scott Hudson has more than 20 years of experience in technical communications, including managing and deploying structured markup and content management systems. He is an active participant in OASIS XML standards committees, including the DocBook, DITA, and Augmented Reality in Information Products Technical Committees. Scott has presented at XML industry conferences. He specializes in information architecture, content analysis, solution design, content model design, XSLT and other XML-related technologies. Currently, Scott is responsible for driving the content strategy for Product Training and Documentation at Jeppesen, a Digital Aviation subsidiary of Boeing.

Email	scott.hudson@jeppesen.com
Website	shudson310.blogspot.com
Twitter	@shudson310
LinkedIn	linkedin.com/in/scott-hudson-01333

Why does a technical communicator need to know this?

Engaging multiple senses helps an audience consume, remember and respond to your content. Some audiences are more visually-oriented, some are more auditory, and some have accessibility needs. Media standards provide a consistent way for you to enhance the communication experience using a variety of the audience's senses.

Sight: Standard image formats include:

- Vector
  - SVG: Scalable Vector Graphics (XML-encoded)
  - CGM: Computer Graphics Metafile
  - U3D: Universal 3D, a format for 3D graphics data
- Raster
  - TIFF: common for high-resolution print and editing
  - PNG: widely used for web, mobile, and print
  - JPG: compressed, smaller files, preferred for photos
  - GIF: widely used bitmap format for web, supports animation
  - BMP: raw, uncompressed bitmap

Standard video formats include: MPEG (MP4), AVI and H.264.

SVG is the only XML-based format for images. It can scale to any display and supports embedded, localized text. Most modern browsers and print engines can render SVG.

Sound: While the video formats support both visuals and sound, standard audio-only formats include: MPEG (MP3), WAV and AIFF. The same markup is typically used to reference audio and video content in XML, though DocBook has distinct elements for `<audioobject>` and `<videoobject>`.

Touch: Primarily found in game content, *haptic feedback* will become more prominent in AR and VR delivery. MPEG-7 provides metadata for describing scenes. More formally, touch-enhanced content can be described using the Haptic Application Meta-Language (HAML) and Sensory Effect Description Language.

Smell and taste: While yet to become mainstream, taste and smell have markup languages under development, but consistent delivery of taste and smell content will require overcoming technical hurdles.

Willam van Weelden
XML Processors

What is it?

Any application that processes XML documents so they can be used by that program or be prepared for use by another program.

Why is it important?

XML is a structured text format. To use XML in technical communication, you need an XML processor for functions such as editing, transforming XML into deliverables, and managing XML in a database.

About Willam van Weelden

Willam van Weelden is a technical writer and functional designer from the Netherlands. After studying to become a teacher, he made the switch to technical documentation.

Willam is an Adobe Community Professional, ranking him among the world's leading experts on RoboHelp. Willam's specialties are HTML5 and RoboHelp automation. Apart from RoboHelp, Willam also has extensive experience with other industry standard programs such as Adobe Captivate and FrameMaker.

Email	contact@wvanweelden.eu
Website	wvanweelden.eu
Twitter	@WvWeelden
LinkedIn	nl.linkedin.com/in/willamvanweelden

Why does a technical communicator need to know this?

An *XML processor* allows you to work with XML documents and their content. Most XML processors are automated tools, part of larger Enterprise Resource Planning (ERP) systems. They perform actions such as transforming content, updating databases, executing work processes, and delivering content to users. Some XML processors are aimed at specific XML actions, such as creating XML-based interfaces between applications. Technical communication requires XML processors that support the following functionality:

- Editing and maintaining XML documents.
- Transforming XML documents into usable formats, such as websites and printed documents. Transforming these documents may require tools that interpret one or more of these languages:
  - **XSLT:** a language for transforming XML documents into other text-based documents, such as another XML format for further processing or HTML for web publishing.
  - **XSL-FO:** an intermediate XML language designed to be transformed into PDF documents. Typically, an XSLT transform converts an XML document into XSL-FO, and a specialized application converts the XSL-FO into PDF.
  - **XPath:** a query language that lets you retrieve content from an XML document.
- Validating the structure of an XML document, guaranteeing its content is complete and contains all required sections. Languages that support validation include: XSD (XML Schema Definition), RNG (RelaxNG), and DTD (Document Type Definition). These languages define the structure of XML documents, including:
  - Tags allowed or required, including their ordering.
  - Attributes allowed or required.
  - Types of content allowed in each tag, such as text, other tags or other content.

For technical communicators, authoring is the most important aspect of using XML. A good XML processor allows technical communicators to focus on content quality while hiding the complexity of XML.

Jane Bozarth
eLearning and mLearning Standards

What is it?

Specifications for eLearning training materials, including file formats, hardware and software requirements, query and statement structures, and protocols for tagging, citing, and attributing material. Examples of eLearning and mLearning standards include the former AICC (Aviation Industry CBT Committee), the current SCORM (Shareable Content Object Reference Model), and the emerging xAPI (the Experience Application Programming Interface).

Why is it important?

By conforming to standards, eLearning content can be created and packaged so as to be shareable across various learning management systems. Standards make eLearning material more available and discoverable, make launch and access more reliable, make tracking and data capture more flexible, and ensure that content is accessible to those with physical or mental challenges

About Jane Bozarth

Jane Bozarth holds a Master's degree in Technology in Training and a Doctorate in Training and Development. Over the past 2 decades she has worked as a classroom designer, trainer, eLearning specialist, and social media specialist.

Dr. Bozarth is the author of many books, including *eLearning Solutions on a Shoestring* and *From Analysis to Evaluation*. She is a popular conference speaker and appears at many industry events.

Email	info@bozarthzone.com
Website	bozarthzone.com
Twitter	@JaneBozarth
LinkedIn	linkedin.com/in/janebozarth
Facebook	facebook.com/Bozarthzone

Why does a technical communicator need to know this?

eLearning and *mLearning* standards were originally developed so that computer-based learning materials could be linked to electronic reporting systems. As the eLearning field has grown, the standards have evolved to support more robust interoperability. With eLearning and mLearning quickly moving beyond traditional courses deployed via desktop or mobile devices, an understanding of standards is critical. Technical communicators developing instructional content – from traditional instruction to video material to serious learning games – will need to understand and adhere to industry specifications.

Learners expect a seamless, unified experience regardless of device or platform. Clients expect content that will run and launch regardless of system. Organizations need reliable data that show performance and increase understanding.

Understanding these standards is vital to ensuring that content is accessible to everyone regardless of ability. A number of authoring, document, and presentation tools now offer help for making material accessible. While there are clearly stated minimum acceptable standards, as the field of *design thinking* evolves, the idea of compassionate design – rather than just designing for compliance – is gaining interest, so a good working knowledge of accessibility is essential.

Finally, an understanding of these specifications helps to ensure that learning material created by technical communicators can be located and accessed across various devices by any learner, with activities tracked and recorded across systems. Increased interest in – and capability for – data capture and reporting that can connect learning activities to job performance makes standards that support interoperability even more critical.

Rebecca Schneider
Terminology Management

What is it?

The practice of actively maintaining dictionaries and glossaries focusing on globally accepted technical standards. Technical terms are organized and controlled, with a clear set of guidelines dictating their use.

Why is it important?

Terminology management enables correct and consistent use of terms throughout the writing process or any other effort requiring accurate vocabulary usage.

About Rebecca Schneider

Rebecca Schneider has extensive background in content strategy, librarianship, knowledge management, and information technology. She has recently formed a new content strategy agency, AvenueCX with Kevin P. Nichols.

She is currently writing a book on the the basics of metadata and taxonomy framework development, focusing on practical, real-world application. Rebecca has worked in a variety of industries, including automotive, semiconductors, telecommunications, retail, financial services, and international development.

Email   rschneider@avenuecx.com
Website   avenuecx.com
Twitter   @1MoreRebecca
LinkedIn   linkedin.com/in/rebeccaschneider/

Why does a technical communicator need to know this?

Terminology (words or phrases) is used to describe how a product works and how people use that product. Terms are also used to tag content to support such activities as reporting and personalization. Technical terms are incorporated into language used in a variety of content types, including product sheets, material declarations, user manuals, knowledgebase support articles, informational videos, and the like.

Content creators choose terminology that reflects their perspective, the perspective of their business unit, and the geographical region. Inconsistent use of terminology can, at the very least, cause confusion or, more seriously, result in legal problems.

The advantages to controlling terminology include the following:

- adherence to technical standards
- linguistic quality
- uniform word usage across an organization (including brand- and company-specific terms)
- conformance to legal requirements
- protection of trademarked and registered product names

Careful terminology management can facilitate translation reuse, reduce errors during authoring or translation, shorten revision time, and help streamline the content creation process.

Terminology management systems maintain terms in a central storehouse and allow for the organization of terms in multiple languages. They manage terms and enable us to provide editorial guidelines based on automated rules. Governance over terms is based both on these systematic rules and on governance committee decisions based on business needs. Leveraging metrics data can also help identify opportunities to improve terminology usage and productivity.

Katherine (Kit) Brown-Hoekstra
Controlled Language

What is it?

A curated set of vocabulary selected to communicate clearly and simply for a specific purpose. Controlled language is often used when writing for machine translation or for global audiences.

Why is it important?

Controlled language is a critical feature of writing for localization. It is an umbrella term that encompasses several initiatives, including Plain Language, Simplified Technical English, and Caterpillar Fundamental English, among others. Effective controlled language initiatives choose the simplest terms needed to convey meaning, while also restricting grammar, syntax, and verb forms.

About Katherine (Kit) Brown-Hoekstra

Katherine (Kit) Brown-Hoekstra is an STC Fellow and STC 2015-16 Society Immediate Past President, an experienced consultant with over 25 years of experience in technical communication, much of it working with localization teams.

As Principal of Comgenesis, LLC, Kit provides consulting and training to her clients on a variety of topics, including controlled language, localization, content strategy, and content management. She speaks at conferences worldwide and publishes regularly in industry magazines. Her blog is www.pangaeapapers.com.

Email	kit.brown@comgenesis.com
Website	comgenesis.com
Twitter	@kitcomgenesis
LinkedIn	linkedin.com/pub/kit-brown-hoekstra/0/321/71b

Why does a technical communicator need to know this?

By including controlled language in your global content strategy, you gain significant benefits:

- **Improved comprehensibility:** controlled language can help make content easier to read and understand.
- **Improved consistency and reuse:** when used with structured authoring (e.g., XML) and component-based content management, controlled language can help you maximize reuse.
- **Better terminology management:** controlled language specifications provide an objective starting point for managing terminology, particularly across disciplines.
- **Improved quality control and efficiency:** by automating tedious tasks like checking compliance, editors can use their time on higher value activities, such as improving the internationalization, organization, and intelligence of the content.
- **Improved quantitative metrics:** several tools exist that enable you to track compliance to a controlled language specification. These tools facilitate benchmarking *before* and *after* editing the content for compliance.
- **Reduced localization costs:** just by limiting vocabulary and reducing word count, companies can save 20% or more on localization costs.

Implementing controlled language is not trivial. When you transition to controlled language, you need to be thoughtful, proactive, and prepared for the long term. Rather than re-invent the wheel, start with the controlled language initiative that most closely matches your needs:

- **Plain Language:** a US government regulation to provide clear communication in government documents. (plainlanguage.gov)
- **Simplified Technical English:** a specification originally developed for the aerospace industry (but now used in many regulated industries) that strictly limits vocabulary and syntax. (asd-ste100.org)
- **Caterpillar Fundamental English:** a vocabulary of about 850 words. Developed in the 1970s, it helped lay the foundation for controlled language.

Controlled language benefits your content strategy by maximizing consistency and reuse, improving efficiency, reducing localization costs, and improving quality control.

Stefan Gentz
European Machinery Directive

What is it?

A directive of the European Parliament and of the Council that applies to machines and manufacturing plants in the European Economic Area (EEA). The purpose of the directive is to promote the free movement of machinery within the Single Market and guarantee a high level of protection for workers and citizens in the European Union.

Why is it important?

Since January 1, 1995 new machines and manufacturing plants may be placed in the European Economic Area (EEA) only if they are compliant with the European Machinery Directive.

About Stefan Gentz

As the Worldwide TechComm Evangelist at Adobe, Stefan's mission is to inspire enterprises and technical writers around the world and show how to create compelling technical communication content with the Adobe Technical Communication Suite of tools. Stefan is a popular speaker at technical communication and translation conferences around the world. He's a certified Quality Management Professional, ISO 9001 / EN 15038 auditor, and Six Sigma Champion. He is also a member of the conference advisory board at tekom / tcworld, the world's largest association for technical communication, and ambassador at the Globalization and Localization Association.

Email	gentz@adobe.com
Website	blogs.adobe.com/techcomm/
Twitter	@stefangentz
LinkedIn	linkedin.com/in/stefangentz
Facebook	facebook.com/stefan.gentz

Why does a technical communicator need to know this?

The European Machinery Directive was first published in 1989. It became mandatory on January 1, 1995. As of 2009, machinery and plant manufacturers must also comply with Machinery Directive 2006/42/EC and the changes implemented with directive 2009/127/EC.

The directive applies to machinery as well as interchangeable equipment, safety components, lifting accessories, chains/ropes/webbing, removable mechanical transmission devices, and partly completed machinery. Some machines, such as tractors, motor vehicles and certain electric and electronic products such as household appliances, office equipment, or weapons have their own directives.

Anyone who wants to sell products or open a manufacturing plant in the European Economic Area (EEA) must keep abreast of this directive – or other European directives that apply to different industries – which includes technical documentation requirements. The manufacturer or its authorized representative must ensure that the machine or plant satisfies the relevant health and safety requirements of the directive and that technical documentation is available.

The technical documentation must include a general description and drawing of the machine, wiring diagrams for the driving circuits, documentation necessary for understanding the machine's functionality, detailed drawings, the risk assessment, the standards applied, and other technical specifications. The documentation must also include all technical reports with the results of exams, a copy of the instruction manual, and a copy of the Certificate of Conformity (EG certificate).

To avoid rework and unpleasant surprises, technical communicators should be aware of these requirements before starting a project.

Deliverable Presentations

Today's audiences demand delivery on many different platforms and devices. This section covers most of the ways technical communicators use to deliver content. Terms range from traditional presentations, such as *print*, to *rich media*, *infographics*, *video*, and *animation*.

Technical communicators have a wide range of presentation vehicles to reach their audiences; a good understanding of the strengths and weaknesses of each, coupled with knowledge of your audience, will help you choose the best deliverable for a given situation.

- HTML5
- Rich Media
- Infographics
- Video
- Animation
- Audio
- eBook
- PDF
- Print

Dave Gash
HTML5

What is it?

The latest version of HTML, the markup language used to structure and present content in pages on the World Wide Web.

Why is it important?

HTML5 is important because it represents a substantial improvement over previous content description languages, including HTML 4.01 and XHTML 1.1. In addition, it provides easier integration with and access to many common, but typically difficult, content presentation technologies and techniques.

About Dave Gash

Dave Gash is the owner of HyperTrain, a Southern California firm specializing in technology consulting and training for hypertext developers. A veteran software professional with over thirty years of development, documentation, and training experience, Dave holds degrees in Business and Computer Science. He is well known in the international technical publication community as an interesting and engaging technical instructor. Dave is a frequent speaker at user assistance conferences in the US and around the world.

Email	dave@davegash.com
Website	davegash.com
LinkedIn	linkedin.com/in/dave-gash-210646b

Why does a technical communicator need to know this?

HTML5 is not a singular thing, it's a collection of discrete features: audio, video, SVG, new elements, new APIs, and much more. Therefore, technical communicators must understand that, although HTML5 is widely supported, support for its features varies across display platforms.

HTML5 simplifies content structural markup through new semantic elements (`<header>`, `<nav>`, `<aside>`). It provides direct access to previously complex non-text content through new media elements (`<audio>`, `<video>`, `<canvas>`). Thus, HTML5 may be considered an enabling technology that makes it easier control the structure and presentation of content regardless of source format, browser version, or delivery device.

All technical communicators should be aware of HTML5's features. Newer versions of many common tools support HTML5 features or transparently provide cross-browser solutions that emulate certain features. HTML5 is not just for the technical communicator who wants to know what's under the hood, but for every technical communicator who wants to stay current.

Matt Sullivan
Rich Media

What is it?
A combination of text, video, audio, and interactive elements designed to convey information and to promote user engagement or interaction.

Why is it important?
Rich media draws the audience closer to the information. It makes content more accessible and increases context, emotional connection, and audience interaction.

About Matt Sullivan
Matt Sullivan helps companies blend text and rich-media content and deliver that content to their audience, often to online and mobile platforms. Matt delivers webinars on behalf of Adobe's Technical Communication Suite team and provides training in TCS applications (specifically, Captivate, FrameMaker, RoboHelp). He also runs courses for the Society for Technical Communication, including training on creating video content for technical communication and Technical Communications Suite applications.

Email   matt@mattrsullivan.com
Website   techcommtools.com
Twitter   @mattrsullivan
LinkedIn   linkedin.com/in/mattrsullivan
Facebook   facebook.com/TC2LS/

Why does a technical communicator need to know this?

As technical communication shifts toward digital delivery, audiences expect content to be easily discovered and accessible on all electronic platforms, including mobile. These platforms provide content delivery options that allow users to craft and direct the experience for themselves.

By recognizing trouble spots within content, technical communicators can identify candidates for rich-media production that will provide more valuable content for both the organization and its audience.

Popular forms of rich media include:

- Camera video
- Recorded computer activities or screencasts
- Software simulations requiring user interaction
- Embedded audio
- Interactive 3D models
- Collections of images or image carousels

Example: Embedded 3D models

A working 3D model can provide an animated, labeled demonstration of how, for example, to disassemble and re-assemble an axle. The rich-media model drastically reduces the text needed to describe the process. At the same time, a 3D model makes it easier to identify and reference the items included with the model. It allows labeling, rotation, and cut-away views at any point in the process.

In the case of embedded 3D, the models that technical communicators need for their content have already been produced by engineers using CAD/CAM software. If so, the 3D model need only be saved in Universal 3D format (U3D) and placed into existing online help or user-guide content.

Text and graphics can't always provide the desired connection with your user. Rich media grabs attention and provides information that linear text can't.

Agnieszka Tkaczyk
Infographics

What is it?

A graphical representation of information that combines text, data visualizations, and images designed to convey the key message in a clear, concise, and enjoyable way.

Why is it important?

In the era of information overload, people want to learn things quickly and effortlessly. Since communicating through pictures is hardwired into our brains, infographics fulfill both conditions.

About Agnieszka Tkaczyk

Agnieszka Tkaczyk is an Information Developer and Enablement Team Lead at IBM. She is responsible for ensuring that users get the right information at the right place and time. Agnieszka is devoted to translating complex ideas and technical jargon into well-structured and user-friendly documentation. She is co-organizer of the soap! conference, the first technical communication conference in Poland.

Email	aga.tkaczyk@soapconf.com
Twitter	@aga_tkaczyk
LinkedIn	pl.linkedin.com/in/agnieszka-tkaczyk-93508570
Facebook	facebook.com/agnieszka.tkaczyk.969

Why does a technical communicator need to know this?

The forgetting curve shows that we forget over 50% of newly-acquired information during the first 24 hours, 80% after a week, and almost everything after a year. The percentages differ from person to person, but the message is clear: we are not particularly efficient at remembering things (Randy Krum, *Cool Infographics: Effective Communication with Data Visualization and Design*).

However, because about 60% of our brain is involved in visual functioning, we are good at recognizing patterns. We are also more likely to better remember information, especially over longer periods of time, if it is accompanied by relevant images. This phenomenon is called the picture superiority effect. Infographics engage this effect to increase the retention of information.

An infographic is like a story. It consists of an introduction, a body, and a conclusion. The introduction must be short enough to be read at one glance and long enough to explain the key message and engage the reader. The body contains most of the information and combines three elements:

- Text
- Data visualizations
- Images that reinforce the message

A good infographic focuses on one key message. The majority of information is presented as data visualizations and images that reduce the amount of time necessary to grasp the key message. Infographics provide technical writers with a visually compelling way of communicating educational or entertaining information in a concise, convenient, easy-to-consume-and-understand package.

Josh Holnagel
Video

What is it?

A series of moving images, typically paired with spoken word, music, or other audio, often for entertainment or educational purposes.

Why is it important?

People turn to video for everything from a quick laugh to learning a new skill. With each passing day it becomes easier to compose and deliver videos, making it a communication form for everyone.

About Josh Holnagel

Josh Holnagel is an instructional designer and graphic artist who specializes in video tutorials. His go-to tool for video creation is TechSmith Camtasia and he's currently most interested in perfecting his voice-over work.

Email	holnagel@gmail.com
Website	holnagel.com
Twitter	@hohoholnagel
LinkedIn	linkedin.com/in/joshua-holnagel-47739352

Why does a technical communicator need to know this?

If you've ever tried to put together a piece of ready-to-assemble furniture or change your own spark plugs, you probably know the frustration that can come from not being able to actually see the hypothetical Tab A inserted into Slot B. When ideas and concepts are difficult to express with words and static images alone, consider using video to get your message across. Video enhances cognitive processing by providing context and detail that is lacking in other media. Video can be optimized for display on a wide variety of device types from large screen projection systems to handheld mobile devices.

Topics like construction and vehicle repair are commonly cited as prime examples for using video, but instructional videos also play a key role in software training and developing workplace and leadership skills. Videos excel at expressing tone, demonstrating complex processes, and presenting abstract concepts.

However, be careful not to go overboard with video. Instructional videos are rarely a 1:1 replacement for guides and help documentation. They aren't easily skimmable or searchable, unless accompanied by a transcription. Burying basic how-to information in a lengthy video tutorial doesn't help someone who only needs to know one specific piece of information. Instead, use video strategically to complement other communication formats or as a powerful means to motivate and inspire.

Bruno Wagner
Animation

What is it?
A moving image that provides visual explanation of a topic or concept that is difficult to explain with words.

Why is it important?
Animation helps people understand concepts that are difficult to explain with words or static images. It reduces complexity for non-expert audiences and shows how single objects or systems move and change.

About Bruno Wagner
Born in 1955, Bruno Wagner has a degree in mechanical engineering. He worked in sales and marketing with SKF/S2M company in the field of mechatronics to promote magnetic bearings and high speed motors. In addition, he researched how to visualize the whole range of SKF products, from giant ball bearings to micro-lubrication devices.

Email   bruno.wagner@kinetikos.fr
Website   kinetikos.fr
LinkedIn   fr.linkedin.com/in/bruno-wagner-28819112

Why does a technical communicator need to know this?

Today, people often expect technical content to contain more than just words. Animation provides images and a visual flow that can increase communication efficiency and help satisfy that expectation.

Animation was originally developed for the entertainment industry, but its scope also includes scientific and technological communication. Animation can be used to convey information rapidly or to support learning in a wide variety of domains, including mechanics, medicine, biology, climate change, and computer science.

Animation is widely used in everything from popular science TV programs to museum presentations to high level visual applications such as flight simulators. It can also be used as part of a conference slideshow or in the body of an online document.

Animation can show objects and situations that would be impossible to observe in reality. Examples include: microscopic parts (very small size scale), climate evolution over many years (very large time scale), hidden objects (internal parts of a machine), and fluids analysis.

Creating scientific and technical animations requires the same skill sets as entertainment animations (from story telling to visual creation). The only difference is the size of the production team: a few experts in science and technology versus a huge entertainment studio.

Depending on the situation, the features of technical communication animation can vary considerably. Examples of the types of variation include the following:

- Schematic or realistic
- 2D or 3D
- Linear or interactive
- With or without sound
- Based on preset images or calculated continuously in real time

Common formats include: animated gif, flash (swf), HTML5, and video formats such as .avi or .mov. Each format, embedded or not, requires its own player.

Ray Gallon
Audio

What is it?
Recorded sound of any type used to transmit or to enhance information.

Why is it important?
We need all of our senses to understand the world. Audio provides an additional channel for learning that helps some users learn faster. It is also vital for people who are blind or visually impaired.

About Ray Gallon
With over 40 years in communication, Ray Gallon is researcher and co-founder at The Transformation Society – a group dedicated to research, training and development around questions of communication, leadership, and complexity. He is also owner of Culturecom, a consultancy specialized in business process improvement through communication. Ray is a frequent speaker at conferences worldwide and a published author. He often presents webinars on technological futures.

Email	infodesign@culturecom.net
Website	transformationsociety.net
Twitter	@RayGallon
LinkedIn	fr.linkedin.com/in/rgallon

Why does a technical communicator need to know this?

You can tell a lot about a car from the sounds the engine makes. The auditory realm includes much vital information that helps us understand the world, including the world of technology.

As technical communicators use more multimedia, sound is becoming as important as text in delivering your message. Some people are visual learners, and others are auditory learners. Including audio, either as an independent element or accompanying images, helps auditory learners master technical products faster.

Sound is also key to helping blind and visually impaired consumers use products. The obvious example of screen readers comes to mind, but sound can also be used to trigger navigation cues to users who cannot see a largely graphical screen. If you combine auditory cues with touch screen technology, a visually impaired person can learn to easily navigate an application. Auditory cues might also be embedded in hardware to identify which part of a device the user is handling.

As new media for communicating technical information develop, the role of audio will evolve considerably. If we take one example, augmented reality, we are often in a hands-free situation where operating a mouse or keyboard or other manual device is inconvenient or even dangerous. Audio information keeps the user focused on the task at hand, without distracting from the essential. In a similar manner, speech recognition software can be used to control the display of information. Technology such as Apple's Siri (Speech Interpretation and Recognition Interface) or Amazon Echo illustrate clearly the accuracy with which this can be done.

Joshua Tallent
eBook

What is it?

A digital file, in one of many possible formats, containing information similar to what would be found in a printed book.

Why is it important?

eBooks allow readers to engage content digitally, with functionality that is not available in print materials. It also allows communicators to reach broader audiences and expand distribution internationally.

About Joshua Tallent

Joshua Tallent, the Director of Outreach and Education at Firebrand Technologies, is dedicated to helping publishers around the world create better books. In addition to heading up training and outreach efforts within Firebrand, Joshua serves on multiple industry committees and working groups and teaches at publishing conferences year-round. He also leads the development of FlightDeck, the most robust EPUB quality assurance tool available.

Email	joshua@firebrandtech.com
Website	firebrandtech.com
Twitter	@jtallent
LinkedIn	linkedin.com/in/joshuatallent
Facebook	facebook.com/FirebrandTechnologies/

Why does a technical communicator need to know this?

An eBook is a digital publication that can be read on computers or other electronic devices. eBooks are typically sold in one of two major formats: ePub (the industry-standard format) and Kindle Format 8 (KF8, formerly known as Mobipocket, Amazon's proprietary ebook format). They can usually be purchased from large retailers like Amazon and Barnes & Noble, as well as from individual publishers and authors.

eBooks ease the distribution of book content around the world by eliminating the need for mailing physical books. They also offer a variety of features that enable readers to more fully engage and benefit from the content, such as searching within the book and following hyperlinks to locations both within the book and outside on the web.

Because eBooks are built in HTML, the same language used to build websites, it is becoming more common to see interactivity, animations, and other feature-rich content in eBook files. eBooks do often suffer from limitations imposed by reading applications and devices (collectively, *reading systems*), but this should become less of an issue as reading systems mature.

The native reflowable nature of eBook formatting makes it easier for readers to engage with content on mobile devices like smart phones. The linking capabilities of HTML can be used to integrate eBook content with other online resources, and the enhancement options make it possible to embed media and other rich content directly inside eBooks. These benefits make eBook formats a preferable alternative to PDF files, which are too limited in their flexibility and functionality.

Maxwell Hoffmann
PDF

What is it?

A file format used to present and exchange documents reliably and without dependence on the operating system, computer hardware, or software.

Why is it important?

Since PDF files can be easily annotated with insertions, deletions, and comments via free PDF Reader software, they are an economical way to manage content reviews from remote staff.

About Maxwell Hoffmann

Maxwell Hoffmann is a content strategist with a strong background in localization, globalization, search engine optimization and technical documentation. He has provided hands-on training to over 1,500 people in scalable, multi-channel publishing solutions. Hoffmann is a periodic author of technical how-to articles for industry journals and is also a frequent presenter at content marketing and intelligent content events. He currently serves as Product Marketing Consultant with MadCap Software and resides in Portland, Oregon

Email	mhoffmann@madcapsoftware.com
Website	madcapsoftware.com/
Twitter	@maxwellhoffmann
LinkedIn	linkedin.com/in/maxwellhoffmann
Facebook	facebook.com/maxwell.hoffmann1

Why does a technical communicator need to know this?

Invented by Adobe, PDF is now an open standard maintained by the International Organization for Standardization (ISO). Although PDF is almost 25 years old, it has emerged as a versatile container for carrying content. PDFs can be password protected to prevent copying or editing, and they may be set to permanently delete sensitive information. All source file information is preserved, even when text, graphics, audio, 3D maps, and more are combined in a single file from multiple sources.

PDF files are searchable, and they are designed to make content accessible to people who are blind or visually impaired. PDF is the most common, standard file format for delivery of patents, court documents, and technical documentation that is under regulatory review.

PDF is not going away any time soon: persistent page and line breaks (possibly with line numbering in the left hand margin) are some of the reasons lawyers demand PDF for documents that may lead to litigation.

PDFs can contain links, buttons, form fields, audio, video, and business logic. More significantly, PDF documents can be signed electronically and are easily viewed using free Adobe Acrobat Reader software. PDFs may be edited, signed, or approved on mobile devices as well as on conventional personal computers.

PDF output is an economical means for creating universal documents that are readable on virtually any device and which support commenting, review, and electronic signatures using free software.

Richard Hamilton
Print

What is it?

Content delivered on paper or other physical media in a form that's readable by humans.

Why is it important?

Some content must be delivered in printed form. Examples include some installation instructions, product packaging, and posters. In addition, many customers prefer to read content in printed form.

About Richard Hamilton

Richard Hamilton is publisher at XML Press. Formerly a manager of technical documentation at Hewlett-Packard, Richard began his career at Bell Laboratories, where he worked on communication software, the UNIX operating system, and internationalization software. He is a member of the DocBook and DITA Technical Committees at OASIS.

He is the author of *Managing Writers: A Real-World Guide to Managing Technical Documentation* (XML Press, 2008).

Email	hamilton@xmlpress.net
Website	xmlpress.net
Twitter	@richardhamilton
LinkedIn	linkedin.com/in/richardlhamilton

Why does a technical communicator need to know this?

Print still holds an important place in most content strategies. Few companies can completely avoid the requirement to create print for at least some content. For example, regulations may require printed documentation, physical products may require printed assembly instructions, and customers may demand print for ergonomic or accessibility reasons.

Print introduces a host of considerations that electronic media eliminate or sidestep, including pagination, margins, hyphenation, and for longer forms, tables of contents and indexes.

Even factors that also apply to the web, such as selecting a typeface, take on a new dimension for print. Fonts that are good, readable choices online can be inappropriate for print, especially at smaller sizes.

You cannot simply generate a PDF from your editing tool and expect it to work well in print. When viewed on line, a PDF can be searched, it can be viewed at different sizes, and it can contain links. Once printed, those capabilities go away, making factors such as fonts, color selection, margins, and indexing more important.

Print poses a different set of accessibility challenges versus electronic media. One challenge is that you cannot rely on metadata or reader-based aids such as adjustable point size, magnification, or screen readers. Instead, you need to think of readable fonts, legible point size, and good color contrast.

Print production has changed significantly in the last five to ten years. *Print on Demand* technology makes it significantly easier and more affordable to create books, pamphlets, and other printed material than ever before. A technical communicator who understands the capabilities of this technology can create high-quality printed content that serves customers and enhances a company's brand.

Future Directions

As with any field, technical communication constantly evolves. This section looks at some of the latest developments in technology that either offer possibilities for delivering information to users or expand the range of technologies that technical communicators may be called on to explain.

Technologies such as *artificial intelligence* and *context sensing* provide new, powerful ways to deliver information. Technologies such as *wearables* or the *Internet of Things* offer new delivery opportunities. They also need talented technical communicators to create content to help people use those technologies.

- Augmented Reality
- Internet of Things
- Artificial Intelligence
- Wearables
- Context Sensing

Christine Perey
Augmented Reality

What is it?

Both the suite of enabling technologies and the resulting experience of a user when highly contextual digital information (text, images, animations, video, 3D model, sound, or haptic stimuli) is presented in a manner that's synchronized in real time with, and appears attached to, physical world people, places, or objects.

Why is it important?

By offering digital information (content) in context and providing interactivity with the content without leaving the physical world, augmented reality profoundly changes how people learn, live, and perform tasks in both physical and digital worlds.

About Christine Perey

Christine Perey is a senior industry analyst and active leader of new technology industry initiatives. In 1991 she saw that it would be possible to improve human communication on personal computing devices with audio and video. She became the editor and publisher of the QuickTime Forum, the first publication for QuickTime developers. She worked as a consultant to the videoconferencing and streaming media industries for over a decade until, in 2006, she realized that the future would lead to augmented reality. She has started and led many communities of interest and currently serves as executive director of the AR for Enterprise Alliance (AREA), the only global member-based organization accelerating AR adoption in the enterprise.

Email	cperey@perey.com
Website	perey.com
Twitter	@cperey
LinkedIn	ch.linkedin.com/in/christineperey

Why does a technical communicator need to know this?

Augmented reality is emerging as the next user interface for presenting and capturing information that pertains to the physical world in context with the user's focus of attention. More than just a novelty or amusement, augmented reality permits users to perceive beyond the superficial and to perform rare and complex tasks with assistance, reducing training and performance times and error rates.

This new approach to presenting, capturing, and interacting with information enables organizations to tap other emerging technology trends such as the *Internet of Things*, cloud computing, high performance and lightweight displays, and high-speed networks to their full potential.

As more people begin to use augmented reality, the role of technical communication professionals will expand with the growing need for information integration into the physical world. The existing professional roles will adapt information authoring, storage, and delivery processes. There will also be new roles created to support the gradual integration of augmented reality software and hardware with connected devices in a variety of different platforms and workflows.

Augmented reality technologies, and the new practices necessary to gracefully deliver and present digital information in context with the physical world, are profoundly changing the human experience.

Mark Lewis
Internet of Things

What is it?

A collection of objects that can communicate and interact with each other sharing data, information, and commands across networks. These objects may be physical devices (virtual or living) and may have the capability to sense and interact with their external environment.

Why is it important?

The Internet of Things (IoT) connects many different types of objects to a variety of technologies that facilitate remote control, diagnostics, and data gathering, storage, and analysis. We can use these capabilities to design new smart products, virtual devices, product clouds, and megasystems to create unprecedented digital ecosystems.

About Mark Lewis

Mark Lewis is the author of *DITA Metrics 101* and is a contributing author of *DITA 101: Fundamentals of the Darwin Information Architecture for Authors and Managers*, second edition. He manages the DITA Metrics LinkedIn group and is a presenter at content industry events. Mark is a Content Engineer and DITA Educator for Quark.

Email	mlewis@ditametrics.com
Website	ditametrics.com/
Twitter	@LewisDITAMetric
LinkedIn	linkedin.com/in/marklewisflorida

Why does a technical communicator need to know this?

The evolution of the Internet from a collection of protocols and billions of web pages to a network that includes an even greater number of interacting objects creates enormous challenges (opportunities). Technical communicators are discovering that each object requires content for users and developers. Some of that content is familiar, for example, programming interface manuals, but some of it will require new types of content.

Many of these objects will be smart objects that include sensors, microprocessors, data storage/analysis/controls, embedded software, and a user interface. Examples include medical devices, smart homes, smart cars, manufacturing machinery, and farming equipment. Smart objects have expanded capabilities that require more marketing, technical, and end-user content.

Smart objects are being combined into systems, such as a smart farm with smart tractors and other connected smart sensing devices. Joy Global has developed a smart mining system and connected several smart mines together.

We have documented systems and systems of systems before. For example, a fleet of ships or a manufacturing plant. However, objects and systems are rapidly becoming smarter. The IoT evolution brings a sharp and unpredictable increase in complexity. Content needs to be embedded in some IoT devices, and some IoT devices will deliver technical content in new, automated ways, using technologies such as augmented reality.

How will technical communicators address these content challenges?

Smart objects are made of smart parts. Smart parts may be software, electrical, or mechanical. Standards, modular content, and intelligent design allow these parts to be reusable, discoverable, reconfigurable, and adaptable. Content is also an important part, so we must use a similar approach and design smart, or intelligent, content that can be reconfigured as needed.

The challenge and opportunity is that the Internet is evolving not just into the Internet of Things but into the Internet of Smart Things (IoST), and content must evolve and become smarter to meet this challenge.

Joe Gollner
Artificial Intelligence

What is it?

A branch of computer science that focuses on the development of software agents, also known as cognitive technologies, capable of performing tasks that would normally require human intelligence, such as finding, interpreting, and manipulating visual and textual information.

Why is it important?

Artificial intelligence is producing cognitive technologies that are radically changing, and even automating, many traditional communication tasks. Technical communicators need to adapt accordingly.

About Joe Gollner

Joe Gollner is the Managing Director of Gnostyx Research, which he founded to help organizations leverage content standards and technologies as the basis of scalable and sustainable content solutions. For over 25 years, he has championed content technologies as an indispensable mechanism to help organizations manage and leverage what they know.

Email   jag@gnostyx.com
Website   gnostyx.com
Twitter   @joegollner
LinkedIn   ca.linkedin.com/in/jgollner
Facebook   facebook.com/joegollner

Why does a technical communicator need to know this?

Artificial intelligence (AI) has been advancing rapidly recently, and, as cognitive technologies, its impact has been spreading. Here are some of the reasons why AI has become so important today including:

- Cognitive technologies have become much more practical, shifting the focus to performing *human tasks* rather than emulating *human thought*.
- Massively scalable big data acquisition, storage, and processing infrastructure has become broadly accessible.
- Decades of research and experimentation in AI, while not successful in emulating human thought, has been successful in improving problem-solving and learning algorithms.

A key area of application for AI is *Natural Language Processing*. Here, tasks commonly performed by people are being increasingly automated, or at the very least facilitated, by intelligent software applications. These tasks include text translation, summation, validation, classification, interpretation, and even generation.

Another area of AI advancement is computer vision, where image and video processing automates the selection, interpretation, and manipulation of visual resources. Yet another important area of AI advancement is information discovery, where contextually aware applications help select relevant information resources for users based on real-time data.

For technical communicators, these changes could not be more significant. More and more traditional communication tasks will be subjected to automated support and even replacement. What this means is that the focus for technical communicators will shift more and more towards the human side of the equation, such as facilitating all-important cross-functional collaborations, the value of which will in fact be increased and not diminished by the advance of AI.

Marta Rauch
Wearables

What is it?
A computing or electronic device that can be worn on your body and that helps you act on information and perform practical tasks.

Why is it important?
With a projected market of $53.2 billion by 2019 (juniperresearch.com), wearables are mobile 2.0. This brings opportunities to technical communicators with expertise in content for wearables.

About Marta Rauch
Marta Rauch is a senior principal information developer at Oracle, where she participates in wearable design challenges and workshops. Marta provides wearable usability feedback and publishes and presents on wearable strategies to worldwide audiences. Marta holds a degree from Stanford University and a certificate from the University of California.

Website	slideshare.net/MartaRauch
Twitter	@martarauch
LinkedIn	linkedin.com/in/martarauch
Facebook	facebook.com/marta.rauch

Why does a technical communicator need to know this?

With the strong market for wearables, technical communicators have opportunities to develop content for wearable devices. It's important to consider wearables as a subject of technical communication and as a content delivery system.

Wearable users have unique content requirements and use cases. For example, you can use an Apple Watch app as a camera remote control when taking photos with your iPhone's camera. Content must be timely and concise on the tiny wearable display.

Practical advice: get involved early in a project. Develop wearable personas and use cases to understand customers' needs. Create content prototypes for wearable interface text and notifications. Participate in usability tests and incorporate feedback.

As a technical communicator for a wearable project, be sure that content meets these guidelines:

- **Useful:** help people complete a task or make their lives simpler, healthier, safer, happier, or more efficient.
- **Timely:** provide messages that are pertinent to the time and place.
- **Unobtrusive:** avoid distracting users with unnecessary content.
- **Relevant:** provide content that pertains to the current task.
- **Concise:** trim content to what is required for the wearable use case.
- **Straightforward:** use a conversational style.
- **Visual:** provide customization instructions; link to videos where it makes sense.
- **Adaptable:** test to confirm that content displays well on the target wearables.
- **Accessible:** ensure that content conforms to accessibility guidelines.

Provide wearable content that meets customer needs to contribute to an excellent user experience.

Christian Glahn
Context Sensing

What is it?
Technologies that infer the characteristics of one or more persons, locations, objects, situations, or activities and use that information to dynamically adapt to, synchronize, and frame situations and processes.

Why is it important?
Context sensing is a key technology for mobile systems, smart objects, and multi-device environments for extending and augmenting human-computer interactions.

About Christian Glahn
Christian Glahn, Ph.D., is professor and the director of the Blended Learning Center at the HTW Chur, Switzerland. Glahn worked in many R&D projects in Austria, The Netherlands, and Switzerland. His research interests are focused around the practices and processes of human learning in complex device ecologies.

Email	christian.glahn@htwchur.ch
Website	blc.htwchur.ch
Twitter	@phish108

Why does a technical communicator need to know this?

Integrating contextual information helps us create and deliver content that better meets user needs. For example, if a user performs a location-based search for nearby restaurants (location context) on a mobile phone, then it is likely that the best matches are those that are open at the time of the search (temporal context) and match the dietary needs of the user (identity context). The context of the user (location, time, dietary needs) is matched to that of the nearby restaurants (location, hours of operation, menu items).

Technical communicators need to understand how to describe these systems, known as context-aware or *smart* technologies, and how to use them for delivering context-aware content that can automatically infer information needs and user intentions.

Context sensing approaches can be categorized by their use of:

- **Predefined information,** including static data such as configurations or manually added information
- **Sensor-based situational information,** such as dynamic data that can be collected directly from hardware sensors connected to an information system
- **Aggregated information,** which integrates and enriches data from sensors or data sources.

Approaches that aggregate, combine, or integrate sensor information are also referred to as *context detection*. Approaches that can relate the contexts of different entities for identifying their situational similarity are called *context matching*. Context detection and matching typically operate on five groups of contextual information for enriching the user experience: individuality, activity, location, time and date, and relationships.

Content Strategy Terms for Technical Communicators

When Ann Rockley wrote her seminal work, *Managing Enterprise Content: A Unified Content Strategy*, in 2001, it seemed obvious to many of us in technical communication that we had been doing some or all of that work for years. Likewise, when it came time to make a list of terms for this publication, the overlap with content strategy was so great that we could have reused most of *The Language of Content Strategy*.

We did reuse a small set of key terms, but many terms we did not reuse are fundamental to both disciplines. Here are some terms from *The Language of Content Strategy* along with some additional terms.

adaptive content
> Content that is designed to adapt to the needs of the customer, not just cosmetically, but also in substance and in capability. Adaptive content automatically responds to the screen size and orientation of any device, but goes further by displaying relevant content that takes full advantage of the capabilities of the device being used.

content engineering
> The application of engineering discipline to the design, acquisition, management, delivery, and use of content and the technologies deployed to support the full content lifecycle.

content lifecycle
> The process that defines the series of changes in the life of any piece of content, including reproduction, from creation onward.

content strategy
> The analysis and planning to develop a repeatable system that governs the management of content throughout the entire content lifecycle.

cutaway view
> A three-dimensional illustration where parts of the exterior are removed to reveal interior features.

design thinking
> A formal method for solving problems that focuses on solutions rather than problems. Traditional problem-solving techniques focus on defining all of the parameters of a problem, then looking for a solution. Design thinking iteratively investigates possible solutions, looking for alternate paths to solve the problem.

document engineering
A methodology for specifying, designing, and deploying the digital documents needed to automate business processes and web services.

Dublin Core
Part of the Dublin Core Metadata Initiative, Dublin Core refers to a set of terms used to describe both digital resources (files, videos, images, web pages, audio, etc.) and physical resources (maps, paintings, sculptures, books, compact discs, etc.).

eLearning
eLearning refers to courses or training delivered using electronic media, most often the Internet.

faceted search
A search technique that allows users to narrow their search by applying filters on multiple, predefined axes. For example, a faceted recipe search might allow searchers to filter recipes based on considerations such as diet (vegan, gluten-free, etc.), main ingredient, cooking time, and level of skill.

forgetting curve
A curve that represents the loss of memory retention over time. First hypothesized by Hermann Ebbinghaus in 1885, the forgetting curve considers the strength of an initial memory and the amount of time that has passed to estimate what percentage of that memory will be retained. Ebbinghaus's experiments showed that retention drops off rapidly over time unless the memory is reinforced.

information architecture
The art and science of structuring information (knowledge) to support findability and usability.

internationalization
A process by which content—online, in print, or in software—is made world ready so it may be localized with minimal rewriting, redesigning, or reengineering.

machine translation
A software-based process that translates content from one language to another.

mLearning
A sub-class of *eLearning* that is delivered on mobile devices.

modular content
A form of structured content that is designed, created, and delivered as discrete components within the content whole.

natural language processing
A field of computer science that marries artificial intelligence and computational linguistics to enable computers to understand the spoken word. It's related to natural language generation: the process of generating the spoken word from a computer.

picture superiority effect
An characteristic of memory that images are more likely to be remembered than words. Infographics take advantage of this effect, using images to help readers retain more information.

print on demand
A printing technology that allows a single deliverable, typically a book, to be printed when needed. With Print on Demand (POD) technology, a publisher does not need to print and store hundreds or even thousands of copies of a book.

progressive disclosure
A technique, often used in user interfaces, that displays only the information a user needs at a given moment, revealing additional information as needed. This technique is designed to simplify user interfaces for users.

style guide
A set of guidelines and standards covering areas such as vocabulary, editing, tone, and voice. May extend to structural aspects of content.

taxonomy
A hierarchical classification scheme made up of categories and subcategories of information plus a controlled vocabulary of terms, usually used to describe a specific area of knowledge.

translation
Conversion of content from one language to another.

translation memory
A repository that contains translated source and destination language pairs.

use by reference
A method for passing information between parts of a system using a link, rather than copying the information. In technical communic-

ation, use-by-reference usually refers to transclusion, where links are used to include content from one source into another.

wizard

An interface that guides a user through a sequence of decisions and forms, helping him or her complete a particular task.

XML

Extensible Markup Language (XML) is an open standard for structured information storage and exchange.

Contributor Index

T

Tallent, Joshua, 104
Tkaczyk, Agnieszka, 96

U

Urbina, Noz, 18

V

van Weelden, Willam, 80
Vang, Erin, 20
Vazquez, Julio, 64

W

Wagner, Bruno, 100
Walsh, Norman, 76
Ward, Christopher, 22
White, Leigh W., 56
Wright, Jan, 46

Subject Index

A

Abel, Scott, 8, 10
accessibility, 38–39, 79, 82–83, 108–109
adaptive content, 123
Adobe, 107
AeroSpace and Defense Industries Association of Europe (ASD), 77
AI (artificial intelligence), 116–117
Amazon, 105
Amazon Echo, 5, 103
analysis, business, 22–23
animation, 100–101
Apple, 103
Apple Watch, 119
architecture
    content, 58–59
    information, 124
    microservice, 68–69
artificial intelligence (AI), 116–117
ASD (AeroSpace and Defense Industries Association of Europe), 77
assistance, user, 16–17, 32–33
audio, 102–103
augmented reality, 112–113
authoring, topic-based, 52–53

B

Barnes and Noble, 105
business analysis, 22–23

C

Caterpillar Fundamental English, 86–87
CCMS, 60–61
component content management system, 60–61, 87
conditional content, 64–67
content, 10–11
    adaptive, 123
    conditional, 64–65

intelligent, 14–15
modular, 14–15, 125
structured, 41, 50–51, 55
content architecture, 36–37, 58–59
content engineering, 123
content lifecycle, 25, 75, 123
    globalization and the, 43
content management, 55, 59, 87
content reuse, 54–57
content strategy, 15, 25, 39, 42, 123
    global, 45, 87
    governance model, 25
    reuse in a, 55
content variables, 66–67
context sensing, 120–121
controlled language, 86–87
customer lifecycle, 74–75
cutaway view, 123

D

data visualization, 97
Day, Don, 8, 50
delivery, dynamic, 53, 61, 67–69
design
    instructional, 32–33
    minimalist information, 72–73
    responsive, 30–31
design thinking, 123
directive, European Machinery, 88–89
DITA, 7
DITA (Darwin Information Typing Architecture), 55, 77
DocBook, 77
document, 12–13
document engineering, 124
DTD, 81
Dublin Core, 27, 124
dynamic delivery, 53, 61, 67–69

E

Ebbinghaus, Hermann, 124

U

Universal 3D format (U3D), 79, 95
usability
    accessibility, 38–39
    defined, 36–37
    user assistance, 16–17
    user experience, 18–19
use by reference, 125
user assistance, 16–17, 32–33
user experience, 18–19, 36–37, 74–75,
112–113, 120–121
    content as extension of, 11

V

validating XML, 81
variables, content, 66–67
video
    animation, 100–101
    defined, 98–99
visualization, data, 97

W

W3C (World Wide Web Consortium),
71
wearables, 118–119
Wikipedia, 53
wizard, 126
World Wide Web Consortium (W3C),
71

X

XML, 126
    media standards for, 78–79
XML document editing standards, 76–77
XML processors, 80–81
XPath, 81
XSD, 81
XSL-FO, 81
XSLT, 81

Colophon

About the Book

This book was authored in expeDITA, a DITA-based wiki developed by Don Day. Contents were converted to DocBook, and the book was generated using the DocBook XML stylesheets with XML Press customizations and, for the print edition, the RenderX XEP formatter.

With the exception of this colophon, the index, and the advertisement at the back of the book, the interior of this book was generated directly from the wiki with no manual intervention.

About the Content Wrangler Content Strategy Book Series

The Content Wrangler Content Strategy Book Series from XML Press provides content professionals with a road map for success. Each volume provides practical advice, best practices, and lessons learned from the most knowledgeable content strategists and technical communicators in the world. Visit the companion website for more information about the series: contentstrategybooks.com.

We are always looking for ideas for new books in the series. If you have any suggestions or would like to propose a book for the series, send email to proposal@xmlpress.net.

About XML Press

XML Press (xmlpress.net) was founded in 2008 to publish content that helps technical communicators be more effective. Our publications support managers, social media practitioners, technical communicators, and content strategists and the engineers who support their efforts.

Our publications are available through most retailers, and discounted pricing is available for volume purchases for educational or promotional use. For more information, send email to orders@xmlpress.net or call us at (970) 231-3624.

The Content Wrangler Content Strategy Book Series

The Language of Content Strategy

Scott Abel and Rahel Anne Bailie

Available Now

Print: $19.95
eBook: $16.95

The Language of Content Strategy is the gateway to a language that describes the world of content strategy. With fifty-two contributors, all known for their depth of knowledge, this set of terms forms the core of an emerging profession and, as a result, helps shape the profession.

Content Audits and Inventories: A Handbook

Paula Ladenburg Land

Available Now

Print: $24.95
eBook: $19.95

Successful content strategy projects start with knowing the quantity, type, and quality of existing assets. Paula Land's new book, *Content Audits and Inventories: A Handbook*, shows you how to begin with an automated inventory, scope and plan an audit, evaluate content against business and user goals, and move forward with actionable insights.

Global Content Strategy: A Primer

Val Swisher

Available Now

Print: $19.95
eBook: $16.95

Nearly every organization must serve its customers around the world. *Global Content Strategy: A Primer* describes how to build a global content strategy that addresses analysis, planning, development, delivery, and consumption of global content that will serve customers wherever they are.

Author Experience: Bridging the gap between people and technology in content management

Rich Yagodich

Available Now

Print: $24.95
eBook: $19.95

Author Experience focuses on the challenges of managing the communication process effectively. It deals with this process from the point of view of those who create and manage content. This book will help you define and implement an author experience that improves quality and efficiency.

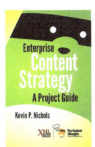

Enterprise Content Strategy: A Project Guide

Kevin P. Nichols

Available Now

Print: $24.95
eBook: $19.95

Kevin P. Nichols' *Enterprise Content Strategy: A Project Guide* outlines best practices for conducting and executing content strategy projects. His book is a step-by-step guide to building an enterprise content strategy for your organization.

Intelligent Content: A Primer

Ann Rockley
Charles Cooper
Scott Abel

Available Soon

Print: $24.95
eBook: $19.95

Intelligent Content: A Primer introduces the concepts, benefits, and building blocks of intelligent content and gives you the information you need to bring this powerful concept into your organization and begin reaping the benefits.

XMLPress.net